Written in an accessible style, this book provides a unique overview of the role and behaviour of soils in both the man-made and the natural environment. The book is structured in two parts: Part A is intended as an introduction to general soil properties and processes, whilst Part B considers soil in relation to the environment, dealing with topics such as the role of soil in supporting plant growth, in maintaining a clean environment, and in the flux of atmospheric gases. Issues such as acidification, contamination with heavy metals, and erosion and conservation are also considered.

SOILS AND THE ENVIRONMENT:
AN INTRODUCTION

SOILS AND THE ENVIRONMENT: AN INTRODUCTION

ALAN WILD

*Emeritus Professor, Department of Soil Science,
The University, Reading*

CAMBRIDGE
UNIVERSITY PRESS

Published by the Press Syndicate of the University of Cambridge
The Pitt Building, Trumpington Street, Cambridge CB2 1RP
40 West 20th Street, New York, NY 10011–4211, USA
10 Stamford Road, Oakleigh, Melbourne 3166, Australia

First published 1993
Reprinted 1994 (twice), 1995

Printed in Great Britain at the University Press, Cambridge

A catalogue record for this book is available from the British Library

Library of Congress cataloguing in publication data

Wild, Alan.
Soils and the environment : an introduction / Alan Wild.
p. cm.
Includes bibliographical references and index.
ISBN 0–521–43280–4. – ISBN 0–521–43859–4 (pbk.)
1. Soils. 2. Soil ecology. I. Title.
S591.W72 1993
631.4 – dc20 92–24680 CIP

ISBN 0 521 43859 4 paperback

WV

Contents

Preface

In writing this book I have tried to show how soils fit into both the natural and the man-made, largely agricultural world. More specifically, my intention has been to help the interested layman and the student understand the behaviour of soils: (i) in supporting the growth of crops, trees and grassland; (ii) in maintaining a clean environment; and (iii) as a source and sink for atmospheric gases. These topics, which have relevance for a wide range of students in earth, biological and environmental sciences, are dealt with in Part B. They are described in simple terms, but knowledge of scientific principles is required if they are to be properly understood; some topics, for example the acidification of soils (Chapter 9), necessarily include an account of chemical reactions. An understanding of soil properties and processes is also needed and the chapters in Part A are intended as an introduction; the serious student is advised to put some effort into reading these chapters. Chapter 1 serves as a general introduction to the text.

Before reading this book two particular points should be noted. First, there are big differences in soil properties between regions of the world, within regions, and often within individual fields. Generalizations about soils can therefore be misleading, as can be the transfer of information between sites or regions unless the relevant soil properties are known and other environmental factors are taken into account. Secondly, environmental issues are often emotive, sometimes being treated unscientifically, or they receive publicity from those who disregard the evidence. These attitudes are to be regretted because the issues, although they can be complex, need to be considered rationally.

Our understanding of the properties of soils, the processes that occur in them and their behaviour under field conditions has advanced greatly in the past few years. It is my hope that the contents of this book will persuade some readers to take the subject further and apply their know-ledge to solving some of the problems that face us.

It is a pleasure to acknowledge the help of colleagues in the Department of Soil Science at Reading who have read, and improved, early drafts of this book, namely Peter Harris, David Jenkinson, Christopher Mott, Stephen Nortcliff, David Rowell, Lester Simmonds, Roger Swift, and Martin Wood; also John Dalrymple, who provided material for the final section of the book. I am especially grateful to Lloyd Jones and Donald Payne for removing ambiguities, correcting errors and improving the presentation, to Keith Shine of the Department of Meteorology at Reading for reading and correcting Chapter 11, to Steve Hopkin of the Department of Pure and Applied Zoology at Reading for supplying some excellent photographs (Chapter 5), and to Kenneth Hassall of the Department of Biochemistry and Physiology at Reading for updating a draft on pesticides in Chapter 13. Any errors that remain are my responsibility. Preparation of a manuscript was made possible by John Wild, who drew several of the figures, and the secretarial staff of the Department, Dorothea Fitzgerald and Sue Hawthorne, who cheerfully re-typed several versions.

February, 1992   *Alan Wild*

Acknowledgements

The author thanks the following for permission to include various Figures and Tables: Academic Press and Dr. Hillel (Figs. 6.3 and 7.3); American Scientist (Fig. 13.5); Australian Institute of Agricultural Science (Fig. 8.5); Blackwell Scientific Publications (Tables 5.2, 6.1 and 7.7); Cambridge University Press (Table 7.1); Elsevier Science (Table 2.7); ICRISAT, India (Fig. 7.5); Intergovernmental Panel on Climate Change (Tables 9.5, 11.1, 11.2, 11.4, 11.5 and Figs. 11.3, 11.6, 11.8); New Phytologist (Fig. 10.1); Oxford University Press (Fig. 11.1); Royal Meteorological Society (Fig. 13.4); Springer-Verlag GmbH (Table 10.2); W.H. Freeman & Co. (Table 6.2); Wiley & Sons (Fig. 13.3, Tables 7.6 and 13.4).

Units, symbols and general information

The units used in the book are either those of the International System (SI) or other metric units where these can be used more conveniently.

SI units

Physical quantity	Name of SI Unit	Symbol for SI unit
length	metre	m
mass	kilogram	kg
time	second	s
thermodynamic temperature	kelvin	K
amount of substance	mole	mol

Each has an agreed definition. The first three units will be generally understood. For our purposes, K equals °C plus 273; one mole of a substance is its molecular mass in grams, e.g. 1 mole of carbon-12 is 12 g, and 1 mole of NaCl is 23 + 35.5 = 58.5 g.

Prefixes

Submultiple	Prefix	Symbol	Multiple	Prefix	Symbol
10^{-1}	deci	d	10	deca	da
10^{-2}	centi	c	10^2	hecto	h
10^{-3}	milli	m	10^3	kilo	k
10^{-6}	micro	μ	10^6	mega	M
10^{-9}	nano	n	10^9	giga	G
10^{-12}	pico	p	10^{12}	tera	T

Other units

Physical quantity	Name of unit	Symbol of unit	Value in SI units
time	day	d	86 400 s
time	year	a	3.15×10^7 s (approx.)
length	centimetre	cm	10^{-2} m
area	hectare	ha	10^4 m^2
volume	litre	l	dm^3 = 10^{-3} m^3
mass	gram	g	10^{-3} kg
mass	tonne	t	Mg = 10^3 kg
pressure[a]	bar	bar	10^5 Pa

[a]See Table 2.2 for units of soil water suction and its equivalent as matric potential.

Indices

Using SI units, length, area and volume are expressed as m, m^2, m^3, respectively. Mass of plant material per unit land area, for example, is given as kg m^{-2} (kilograms per square metre), or often more conveniently as kg ha^{-1} (kilograms per hectare) or t ha^{-1} (tonnes per hectare). A rate process, for example the rate of photosynthesis per unit land area, is expressed as kg m^{-2} s^{-1} (kilograms per square metre per second) or as kg m^{-2} d^{-1} (kilograms per square metre per day) or as t ha^{-1} a^{-1} (tonnes per hectare per year).

Concentrations

Various expressions are used. Concentrations of salts, acids and alkalis in solution are usually expressed as mol l^{-1} (moles per litre), which is equal to mol dm^{-3} (moles per cubic decimetre), and has the symbol M (molar) As an example, a solution of sodium chloride (molecular mass 58.5) containing 5.85 grams per litre has a concentration of 0.1 mol l^{-1}, or 0.1 M.

Also used are g l^{-1} (grams per litre) or submultiples of these units, e.g. mg l^{-1} (milligrams per litre) or μg l^{-1} (micrograms per litre). Small concentrations are often expressed as ppm (parts per million, 10^6), ppb (parts per billion, 10^9) or ppt (parts per trillion, 10^{12}). As an example, 1 g of sodium in 10^6 g water (equal to 1 mg l^{-1}) has a concentration of 1 ppm. Gas concentrations are often expressed as ppmv (parts per million by volume); for example, a concentration of 350 ppmv of CO_2 in air is 350 m^3 of CO_2 in 10^6 m^3 of air. If it is not clear whether volume or mass is being used, ppm and similar expressions are best avoided.

Elements

The following is a list of chemical elements to which reference is commonly made in the text.

Element	Symbol	Atomic mass	Element	Symbol	Atomic mass
Hydrogen	H	1.0	Manganese	Mn	54.9
Boron	B	10.8	Iron	Fe	55.8
Carbon	C	12.0	Cobalt	Co	58.9
Nitrogen	N	14.0	Nickel	Ni	58.7
Oxygen	O	16.0	Copper	Cu	63.5
Fluorine	F	19.0	Zinc	Zn	65.4
Sodium	Na	23.0	Arsenic	As	74.9
Magnesium	Mg	24.3	Selenium	Se	79.0
Aluminium	Al	27.0	Bromine	Br	79.9
Silicon	Si	28.1	Strontium	Sr	87.6
Phosphorus	P	31.0	Molybdenum	Mo	95.9
Sulphur	S	32.1	Cadmium	Cd	112.4
Chlorine	Cl	35.5	Caesium	Cs	132.9
Potassium	K	39.1	Mercury	Hg	200.6
Calcium	Ca	40.1	Lead	Pb	207.2
Chromium	Cr	52.0			

1

Introduction: soil in a natural and man-made environment

Soil is one of the familiar materials that we take for granted. We grow plants in it and it sticks to our boots when we walk over it. Usually our interest is utilitarian, and rightly so. We might ask what can be done to improve the growth of plants, or how a good surface for games can be maintained. These sorts of questions can be answered, although we are not always aware of what is possible (and the remedy might be too expensive).

Important for the survival for the human race is the support that soil provides for the growth of arable crops, grassland and trees, which produce food, fibre for clothes, and timber for buildings and fuel. Together with water, air and radiation from the sun we depend on the half metre or so of the Earth's crust to provide these essentials for life. The general properties of soil that make it useful are those listed in Table 1.1.

Soils and food production

In Europe since Roman times, and probably much earlier, there has been awareness of techniques to maintain soil fertility. In South and Central America and in south-east Asia, terraces were built several centuries ago to conserve soil against erosion in order to ensure that food crops could be grown. In many parts of the world some form of

1

Table 1.1. *Properties of soil that make it useful*

1. Provides water, nutrients and anchorage for plants and trees in natural forests and grasslands, annual and perennial crops and planted grassland.
2. Provides the habitat for decomposer organisms which have an essential role in the cycling of carbon and mineral nutrients.
3. Acts as a buffer for temperature change and for the flow of water between the atmosphere and ground water.
4. Because of its ion exchange properties it acts as a pH buffer, and retains nutrient and other elements against loss by leaching and volatilization.

fallow, whether grassland or woodland, has provided a rest period for the recovery of soil fertility. These examples show that farmers have regarded soil as a resource on which their livelihood depended, although mistakes have been made as will be referred to later.

The increasing human population of the world, particularly in the Third World, presents a new problem. Whereas in the past more food could be grown by cultivating new land that required few purchased inputs, the opening up of the prairies in North America being the best known example, most of the land that remains is either unsuitable for development or requires inputs that can be expensive, including fertilizers, soil conservation measures and irrigation.

Soils and the environment

In Europe, North America and in a small number of countries elsewhere, the development of higher-yielding crop varieties and more intensive use of fertilizers, pesticides and irrigation have led to the overproduction of food, including animal products. The call is now for less intensive practices. Also, there is public concern about the quality of food, drinking water and air. This concern has led to criticism of the use of fertilizers and pesticides, and we might ask whether this concern is justified.

These questions relate to the effects of crop production on the human environment and are discussed in later chapters. What should also be recognized is that soils are part of the natural environment. They partly determine the distribution of plant species, provide a habitat for a wide range of organisms, buffer the flow of water and solutes between the atmosphere and ground and surface waters, and act as both source and sink for gases in the atmosphere.

Investigation of soils can therefore be directed at problems to do with growing crops, such as removing limitations to the achievement of optimum yields, or the prevention of harm to the environment. Alternatively, soils can be considered as a component of the natural global system, and this will be described first.

1.1 Some definitions

Like all scientific subjects, soil science has its own terminology. A few definitions are needed at the outset and others are given later.

Soil Unless stated otherwise, soil in this text refers to the loose material composed of weathered rock and other minerals, and also partly decayed organic matter, that covers large parts of the land surface of the Earth. As used in an agricultural context, soil supports crop growth and can be tilled. The word has a different meaning when applied to the unconsolidated material on planets which contains no evidence of biological activity at present or in the past.

In describing soils two further definitions are required.

A soil profile is defined as the vertical face of a soil that can be exposed, for example, by digging a pit or in a road cutting. It includes all the layers (horizons) from the surface down to the parent material (Figure 1.1). Within the soil profile the part that contains plant roots or

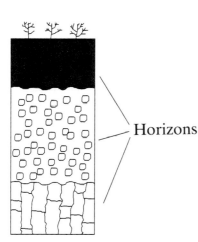 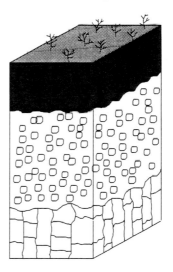

Soil profile   Pedon: three-dimensional

Figure 1.1 Schematic representation of a soil profile and a pedon.

is influenced by plant roots is called the solum. It is this part which is examined in the field by soil surveyors.

A pedon is the smallest volume that can be called a soil. The point of this definition is that a soil is three-dimensional, that is, it has lateral extension as well as the two dimensions seen in a vertical face. A pedon is therefore a vertical slice of a soil profile of sufficient thickness and width to include all the features that characterize each horizon.

The processes by which soil profiles are formed, and the wide range that occurs over the Earth's surface, are discussed in Chapter 3.

1.2 Soil as a component of ecosystems

An ecosystem (or ecological system) can be defined as a community of interacting organisms and its environment functioning as a reasonably self-sufficient unit. As illustrated in Figure 1.2, a terrestrial ecosystem consists of primary producers (trees, herbs, grasses), and decomposers (microorganisms, herbivores, carnivores). The primary producers are photoautotrophic, that is, they use some of the energy from the sun to convert atmospheric carbon dioxide into organic compounds, a process that requires water and nutrients which are supplied from the soil. The organic compounds are used by herbivores and carnivores as a source of

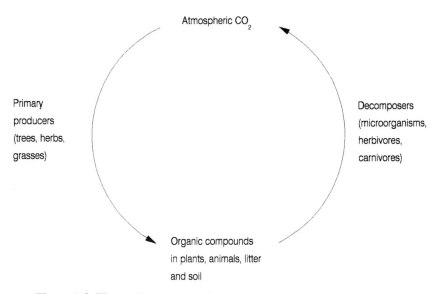

Figure 1.2 The carbon cycle in the atmosphere–plant–soil ecosystem.

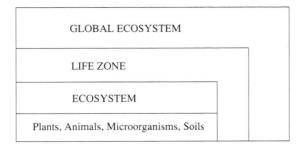

Figure 1.3 Soils, plants, animals and microorganisms form an ecosystem such as a tropical forest; all the world's tropical forests constitute a life zone and all the life zones (grasslands, other forests, tundra, oceans etc.) form the global ecosystem.

energy, and carbon compounds are used to build their tissues; some of the energy is lost as heat and some of the carbon is respired as carbon dioxide.

The most active group of decomposers are the soil microorganisms. By processes described in Chapter 5, the end product of oxidation of carbon in organic compounds is carbon dioxide, which returns to the atmosphere. The nutrients taken up from the soil are partly retained in the vegetation and the animals, and partly returned to the soil. Although there are additions from outside, especially of water, and there are usually leakages, for example in drainage water, the soil, vegetation and associated animals form a unit which is roughly self-contained.

The unit comprising an ecosystem can be of almost any size (Figure 1.3). At one extreme, planet Earth can be taken as the unit and solar radiation is then the only external input. At the other extreme, a soil forms an ecosystem, its organisms being members of a community which interact with each other and with their physical and chemical environment. The inputs are carbon compounds from the primary producers, water, oxygen and nitrogen from the atmosphere, and essential nutrients from mineral weathering. Between these extremes are life zones, in which soils support a particular type of flora and fauna and there is a characteristic climate. Tropical rainforest and boreal forests are examples of life zones.

1.3 Soils through geological time

By the definition given above, soil must have existed from the time that the land was first colonized by living organisms, probably photosynthetic bacteria. Multicellular land plants appeared in the Ordovician period about 450 million years ago, and trees had evolved by the middle Devonian period (about 370 million years ago).

Soils similar to those of the present day have been found in sedimentary rocks dating from the Devonian and later geological periods. Two examples from the Carboniferous period of 300 million years ago are shown in Figure 1.4. Both profiles contain structures that can be identified as fossilized roots and a surface organic layer, O, that is now coal.

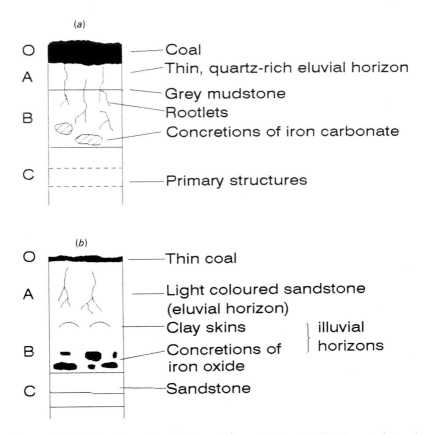

Figure 1.4 Fossil soil profiles (palaeosols) containing fossilized roots from the Carboniferous period (300 million years ago); profiles range from 0.3 m to over 1.5 m in thickness. (From P. Wright, Postgraduate Research Institute of Sedimentology, University of Reading.)

Profile (*a*) shows evidence of chemical reduction of iron and its removal from the bottom of the A horizon; profile (*b*) contains an A horizon from which clays and iron have been leached out, and a lower horizon, B, where they have accumulated. The C horizon refers to the underlying material, which is unaffected by soil-forming processes. These features can be seen in soils forming at present, as discussed in Chapter 3.

During the past one million years large areas of land in Europe, Central Asia, China, parts of the USA and elsewhere in the world have been covered with deposits of loess (fine-grained material of wind-blown origin). At some sites the loess buried a complete soil, including the surface, organic horizon; at others, deposition was on top of a partly eroded soil. Cores of loess in Europe and Asia have revealed up to 40 buried profiles in a vertical section, indicating soil formation between periods of loess deposition.

The properties of soil are determined by climate, organisms (including man), relief, parent material and time, as discussed in Chapter 3. As we increase our understanding of the relation between a soil property and these factors, so can we deduce with increasing confidence the conditions, for example climate, under which buried soils were formed.

1.4 Soils and man

The past

Cultivation of the soil to grow wheat, barley and other crops probably originated about 10 000 years before the present (10 000 BP) in the valleys of the northern, upland area of Mesopotamia, now Iraq. The soil was a friable silt loam. Two thousand years later there were villages in the southern valleys of the Tigris and Euphrates and crops were grown under irrigation on the alluvial soils. The production of food was sufficient to support the populations of large cities for several centuries during the great civilizations of Sumer, Akkad and others.

It is not known for certain what ended the Mesopotamian civilizations. Wars were a factor; it is also known that villages were devastated by floods, irrigation channels became filled with sediment, and soils in the upland areas became gullied. It is therefore possible that erosion of soils in the uplands, although initially providing fertile alluvial soil in the valleys, contributed to the loss of food production and to the decline of the civilizations. If this is so, it is the first known example of man-made soil erosion affecting a whole watershed; there have been many since.

Increasing soil salinity might have been another cause of reduced food production.

Other early civilizations developed in the valleys of the Nile, the Indus and the rivers of China. Frequent floods from the rivers provided water and silt, and the silt provided plant nutrients. Crops that were grown included beans and vetches which, as legumes, provided their own nitrogen by fixation from the air. These valley civilizations therefore had the basis for long-term, and possibly permanent, crop production.

Cultivation of upland soils and overgrazing of herbage in Europe, South and Central America and south-east Asia occurred as populations increased. The mature trees were felled for timber and fuel and on steep slopes, as in the mountains of Greece, soil erosion occurred. Loss of soil and the consequential loss of crop production is believed to be one reason for the colonization of other lands by people in the Mediterranean region in order to grow food for the home country. Terracing to conserve soil against erosion was, however, practised by some peoples such as the Incas in Peru.

The history of soil erosion has continued to the present day. Drought in North America and Australia in the 1930s led to severe wind erosion, and the recent drought has resulted in wind erosion in the African Sahel. In all these areas the erosion was exacerbated by cultivation and overgrazing. All countries have soil erosion to a greater or lesser extent, a problem that is discussed in Chapter 12.

Erosion is the most serious form of soil degradation because it cannot be reversed. Other forms are acidification, nutrient depletion and accumulation of salts, each of which is capable of being reversed. These problems are dealt with in later chapters. All arise from more intensive use of soils, and often those which should be left undisturbed. As will be discussed next there is a rapidly increasing population to be fed, which will depend on the maintenance of soil fertility.

The future

The total population of the world has more than doubled since 1950 (Figure 1.5). It is expected to exceed 6 billion by the year 2000 and might reach 10 billion before the year 2100. The rate of growth of populations is greater in poorer countries than in richer, using the gross national product (GNP) as a measure of wealth (Table 1.2). The requirement for food will rise correspondingly.

Table 1.2. *Population increase and gross national product (GNP) in selected countries during the period 1986–89*

Country	Population Number (millions)	Population Annual increase (%)	GNP per capita (US $)
Tanzania	25.2	3.3	258
India	833.4	2.0	290
China	1112.0	1.6	320
Kenya	24.3	4.2	370
Ghana	14.8	2.9	410
Malaysia	16.7	2.0	2 092
Brazil	150.8	2.0	2 130
Greece	10.0	0.3	4 670
UK	57.0	0.2	13 329
France	56.0	0.3	16 800
USA	248.2	0.9	19 800

Source: From *Handbook of Nations*, Gale Research Inc., Detroit, 9th edition.

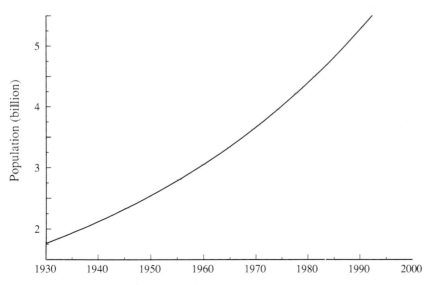

Figure 1.5 The human population of the world since 1930 (FAO Production Yearbooks).

Table 1.3. *Use of land in the world*

Use	Area (10^9 ha)
Arable and permanent crops	1.48
Permanent pasture	3.21
Forest and woodland	4.05
Other land	4.33
Total	13.07

Source: FAO Production Yearbook, 1989.

Extra food can be produced by bringing new land into cultivation, by more intensive use of the existing cultivated land or by a combination of the two. According to the figures in Table 1.3, about 1.5 billion hectares (11.3% of the total land surface) are used for arable and permanent crops. Estimates of the remaining area that could be developed for arable cropping depend on the practicability of providing fertilizers, lime, soil and water conservation measures, and irrigation. Clearly there will be economic constraints, and this makes it difficult to give a realistic estimate. Further, the land that could be developed is unevenly distributed, there being most in South America and Sub-Saharan Africa and relatively little elsewhere. Land is also needed to provide wood for timber and fuel, for domestic animals, and to be preserved in natural ecosystems.

The Green Revolution, which included the introduction of short-stemmed varieties of cereal crops, the use of fertilizers, pesticides and often of irrigation, has shown that crop yields can be several-fold higher than those of the subsistence farmer. Improved crop varieties are continually being introduced, some of which are suitable for short growing seasons, and others for acid, saline, or nutrient-poor soils. With these and other developments, estimates of potential food production of the world indicate that the projected population can be fed, but for the requirement to be met it will be necessary to manage soils better than in the past.

1.5 Soils and pollution

Two definitions are needed.

Pollution occurs when some part of the environment is made harmful or offensive to organisms and especially to humans.

Contamination is the addition of a hazardous chemical or organism where no harm has been demonstrated.

In both definitions the interests of humans are regarded as paramount. For example, an insecticide that is used to kill locusts is not regarded as a pollutant unless it also kills other organisms that we consider to be beneficial or wish not to harm. When the insecticide falls on soil it contaminates it, and it becomes a soil pollutant if it harms beneficial soil organisms. In common usage 'pollution' and 'contamination' are often used synonymously.

All chemicals are harmful when their concentrations are high enough even though they are harmless or even beneficial at a low concentration. A good example is zinc, which is an essential element for plants and animals but is harmful to both at high concentrations.

Two kinds of process in soil generally prevent the accumulation of chemicals at concentrations which are harmful: decomposition of organic chemicals, including pesticides, by microorganisms; and adsorption and precipitation reactions on soil components. These processes are discussed in later chapters.

The maximum permissible concentrations of many chemicals in soil, air, water and food are now the subject of legislation. As a concentration lethal to humans cannot be established by experiment there will always be uncertainty as to what concentrations are safe. The general principle is to protect infants, the old and the sick, who are most at risk. Permissible concentrations of heavy metals in soil, and nitrate and pesticides in drinking water are referred to in Chapters 10 and 13, respectively.

Pollution by natural causes

Most attention is given to pollution caused by human activities. Soils, food and water can, however, have toxic concentrations of elements owing to high concentrations in soil parent material or emissions from volcanoes. Two examples will be given.

High concentrations of selenium occur in certain plant species of the genus *Astralagus* growing on soils derived from Permo-Cretaceous sediments in the Plains and Rocky Mountains of the USA. Animals grazing these plants suffer from loss of hair and hooves. Another problem wth grazing animals, especially cattle and sheep, is due to high concentrations of molybdenum in so-called 'teart' pastures. These occur in the west of England and in Ireland on imperfectly drained clay soils with high organic matter contents derived from Lower Lias clay. The same problem occurs in the western states of the USA on poorly drained clay loams with high contents of organic matter, which are derived from granitic alluvium.

This chapter serves as a general introduction. The next five chapters give an outline of the main properties of soils that are referred to later in the book. They include descriptions of the properties of soils and the processes that occur in them. They show that soils are populated by vast numbers of living organisms and contain components with important but rather unusual physical and chemical properties, and they indicate the reasons why soils vary over the surface of the Earth. The chapters are included to provide a reasonably firm base on which to build an understanding of the environmental effects of soils.

1.6 Summary

Soil is an essential component of the terrestrial ecosystems of the Earth. It supports plant growth and provides a habitat for large numbers of animals and microorganisms that decompose leaf litter and plant residues, thereby helping to cycle the nutrients on which plant growth depends. It has had these functions since plants first colonized the land.

Soil also supports the growth of arable crops, grassland and trees on which man depends for food, fibre, and wood for fuel and as a building material. An increasing world population requires more of these resources. This requirement can be met by bringing more land (much of which is unsuitable) into cultivation, by more intensive use of land, or by a combination of the two. Each can cause problems of soil degradation and pollution. These problems are discussed more fully in the chapters in Part B, to which the reader familiar with the properties of soils may wish to turn.

A
Soil properties and processes

2

The soil components

2.1 Introduction

Examination by eye of a spadeful of soil may show the following:

aggregates of particles a few millimetres or centimetres in size which, when wet, can be crushed by the fingers;
roots of plants;
pieces of dead roots, stems and leaves, which may be partly decayed;
stones and gravel (greater than 2 mm diameter) and sand particles (0.02 to 2 mm diameter);
earthworms and various arthropods, including insects;
spaces between the solid material occupied by air or water or both.

Under an optical microscope smaller particles of silt (0.002 to 0.02 mm diameter) can be observed, and the higher power of an electron microscope will reveal the presence of clay particles (less than 0.002 mm diameter). The dark colour of soil shows the presence of humus. From

15

each spoonful of soil millions of cells of bacteria and other microorganisms can be isolated. If a bottle is half-filled with damp soil and then stoppered, the composition of the air above the soil will change, oxygen being replaced by carbon dioxide due to respiration by soil organisms.

From the above introduction it will be seen that soil contains mineral and organic material, air, water and living organisms. The properties of a particular soil depend largely on the proportion and composition of these components and how they interact with each other. The properties may, however, change because soil is exposed to the weather, plants grow on it and die, it is trampled on by animals, and man cultivates crops on it.

In order to understand the properties of soil and how they can be changed we must first consider the components themselves. In this chapter we describe the mineral and organic components, air and water. Soil organisms are described in Chapter 5.

2.2 Texture and structure

Soil texture

This refers to the relative proportions of clay, silt and sand in a sample of soil. The dominant size fraction is used to describe the texture, for example, as clay, sandy clay, silty clay etc. If no fraction is dominant the soil is described as a loam. A triangular diagram is commonly used to set the limits for each texture class (Figure 2.1). In this figure there are three particle sizes: sand, silt and clay; each is expressed as the percentage of the soil which passes through a 2 mm sieve (the fine earth). If coarser material is present in substantial amounts an additional description such as stony or gravelly can be included. The sand fraction may be subdivided according to particle size (Table 2.1).

This method of describing texture requires knowledge of the particle size distribution, as obtained by sieving, and using the velocity of sedimentation under gravity for fine fractions (for methods, see the suggestions for further reading at the end of the book). With experience, the textural class can be recognized by rubbing moist soil between the fingers. Coarse sand feels gritty, fine sand feels silky, silt gives a smooth and non-sticky feel, and clay sticks to the fingers. This is a quick method of assessing texture and is especially useful to the soil surveyor in the field.

Texture is an indicator of other soil properties, but used alone it has

Table 2.1. *The classes used to describe the distribution of size of soil particles (mm diameter)*

	International system	USDA system[a]
Gravel	above 2.0	above 2.0
Very coarse sand	—	1.0–2.0
Coarse sand	0.2–2.0	0.5–1.0
Medium sand	—	0.1–0.5
Fine sand	0.02–0.2	0.05–0.1
Silt	0.002–0.02	0.002–0.05
Clay	less than 0.002	less than 0.002

[a]This system of the United States Department of Agriculture is also used in several other countries.

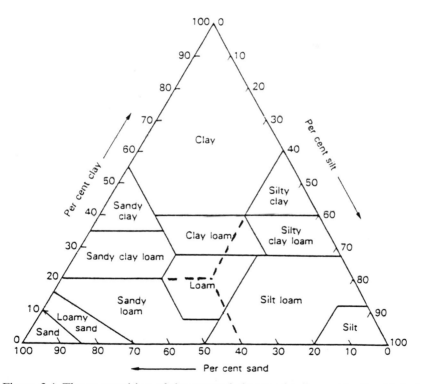

Figure 2.1 The composition of the textural classes of soils used by the United States Soil Survey. (Sand, 2–0.05 mm; silt, 0.05–0.002 mm; clay, less than 0.002 mm.) A soil with 40% sand, 40% silt and 20% clay is described as a loam, as shown by the broken line.

limited predictive value of these other properties. For example, the ability of a soil to adsorb cations from solution depends on the mineralogy of the clay fraction as well as on the percentage of clay. It also depends on the amount and nature of the organic matter the soil contains. The permeability of soils to water depends more on the organization of the mineral particles and organic matter into structural units with pore spaces between them than on texture itself. Texture does, however, indicate the ease with which a soil may be cultivated. Soils high in clay are often described as 'heavy' because they require high power for cultivation, whereas sandy soils are known as 'light'. Also, clay soils retain more water against gravity than sands and consequently warm up more slowly in spring (Section 6.9).

Soil structure

This describes the arrangement of the soil particles. A soil in which these are unattached to one another is said to have single-grained structure, or be structureless; this occurs with coarse-grained materials, as in sand dunes. At the other extreme is massive structure where all the mineral particles are packed tightly together, as occurs in some clay soils. More usually the particles form aggregates, which have a size and shape that are often characteristic of the soil. Aggregates formed by natural processes are known as peds.

For arable cropping a soil should consist mainly of small aggregates or crumbs, which allow seedlings to emerge easily and provide a ready supply of water and oxygen to plant roots. This desirable state is brought about by judicious cultivations and is described as tilth. The aggregates should be sufficiently strong to resist being broken down by the impact of raindrops, otherwise the soil surface may form a crust which can prevent seedling emergence and may also lead to run-off and erosion. Size, stability and internal porosity are the important properties of aggregates.

Clay particles stick to one another in most soils, the exception being 'alkali' soils in which sodium ions cause the clays to disperse. In all other soils the clay is flocculated (the 'plate-like' particles stick to one another) and the clay particles may be stacked face to face in packets that are sometimes known as domains. These domains may be linked together through organic molecules to form microaggregates, several of which form a visible aggregate that may incorporate sand and silt particles. The most stable aggregates are formed in neutral and calcareous soils (containing calcium carbonate) under grassland in the presence of sufficient clay, and also in certain tropical soils in which the primary mineral particles are cemented together by iron oxide.

Organic matter is an important component of aggregates. Work in Australia showed that fungal hyphae and fine roots bind together aggregates larger than 2 mm. Medium sized aggregates (20–250 μm diameter) were more stable because of binding by humus, iron and aluminium oxides and clay particles. Smaller aggregates were yet more stable but contained the same binding agents. The effect of humus probably depends on its content of microbial polysaccharides that contain carboxyl groups (Section 2.8).

2.3 Bulk density and pore space

The density of the soil mineral particles is usually from 2.6 to 2.7 g cm^{-3} (2600–2700 kg m^{-3}). When dry the bulk density of soil is often about half this value because the voids between the particles are filled with air.

The bulk density (more properly, the dry bulk density) is defined as the ratio of the mass, M, of dry soil to its volume, V. It is given the symbol ρ_b, where $\rho_b = M/V$ (ρ is Greek rho). Because the value of ρ_b is of most relevance to the behaviour of soil under field conditions it is measured in the field. This is usually done by driving an open-ended cylinder into the soil, smoothing the ends and then finding the dry mass of the enclosed soil. It is a variable property because of the effects of weather, cultivation and compression by animals; it varies over small distances and generally increases with depth in the profile. Commonly, ρ_b has a value between 1.0 and 1.6 g cm^{-3}.

The fractional soil pore space, e, also known as the soil porosity, is calculated from the dry bulk density, ρ_b, and the particle density, ρ_s, as

$$e = 1 - (\rho_b/\rho_s),$$

where ρ_s is usually between 2.6 and 2.7 g cm^{-3}. The percentage of pore space in the soil volume is:

$$\% \text{ pore space} = 100[1 - (\rho_b/\rho_s)]. \tag{2.1}$$

The pore space is occupied by water and air. If θ (Greek theta) is the volume of water per unit volume of soil, the fractional air volume e_a, is ($e - \theta$). Under conditions that support good growth of plants, e_a and θ are about equal, so if $\rho_b = 1.3$ g cm^{-3} and $\rho_s = 2.6$ g cm^{-3}, $e = 0.5$, and e_a and θ should both be about 0.25. Taking the density of water equal to

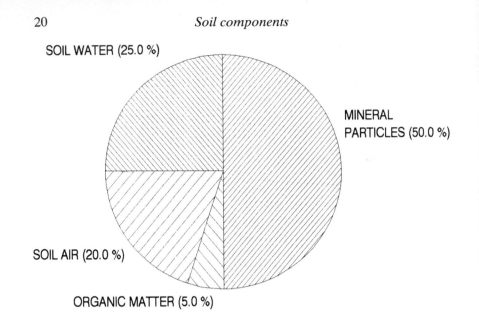

SOIL WATER (25.0 %)

MINERAL
PARTICLES (50.0 %)

SOIL AIR (20.0 %)

ORGANIC MATTER (5.0 %)

Figure 2.2 The soil components include mineral particles, organic matter, water and air. The percentage of each component shown in the figure is the percentage of the soil volume, and each is typical of the top 10–15 cm of a soil. The total pore space in this example is 45%.

1 g cm^{-3}, a value of $\theta = 0.25$ means that each cubic centimetre of soil contains 0.25 g water. The relationship between the volume of soil occupied by solid matter, water and air is shown in Figure 2.2.

Soil bulk density is needed when converting the analysis of a known mass of soil to a volume or area of soil. If, for example, the soil contains w g of water per gram of soil, the volumetric content of water (density of 1 g cm^{-3}) per cubic centimetre of soil is $w\rho_b$ cm^3. The mass of water to 15 cm depth in 1 m^2 of soil would be $15 \times 10^4 \, w\rho_b$ grams.

Because of the presence of aggregates and microaggregates, the total pore space, e, consists of spaces differing in size. If the soil profile is free-draining, water drains out of the large pores, which are then occupied by air, and water remains only in the small spaces and pores (Figure 2.3). The supply of water and air to plant roots and soil organisms therefore depends on the total pore space, on the sizes of the pores and on the drainage conditions of the site. The pores have irregular shapes but it is useful to classify them in terms of effective diameters as though they were cylindrical tubes:

Transmission pores (they conduct water rapidly
through the soil)   >50 µm

Mineral particle

Water

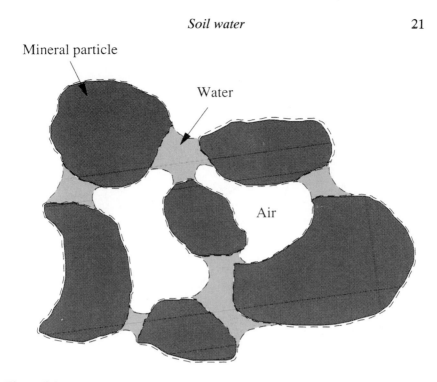

Air

Figure 2.3 Water, air and mineral particles in soil. In a freely draining soil, air occupies the large pore spaces. Water is present in the smaller pore spaces and as films on soil particles; it is also present within aggregates (not shown).

Storage pores (they retain water against gravity)   0.5–50 μm
Residual pores (they remain filled with water
even when the soil appears to be dry)   <0.5 μm

Pores larger than 50 μm (0.05 mm) will drain under gravity. Wide pores up to several millimetres across may exist as cracks in clay soils, and as channels created by burrowing animals, for example earthworms, or left by decayed roots.

2.4 Soil water

This section deals with the liquid phase in soil. Water vapour is referred to in Section 2.6 and the solids and gases dissolved in the water are discussed in Section 2.5.

A sample of coarsely structured soil which is saturated with water will lose water if it is allowed to drain. More water is lost if pressure is applied to the top of the soil or suction is applied to the bottom. The loss

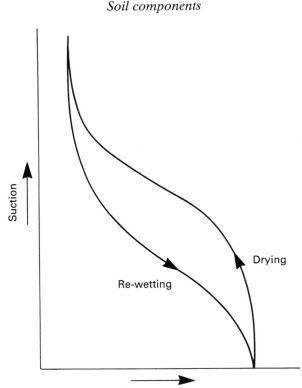

Figure 2.4 The water content of a soil sample drying under suction and then re-wetting.

increases as the pressure or suction is increased. The relationship between water content and suction is shown in Figure 2.4. This shows a different relationship for a soil being dried by increasing suction and one that is being wetted, and a few words of explanation are needed.

Because soil pores are irregular in shape the suction required to *remove* water is determined by the narrow parts of the pores where the water is most strongly held; water will then empty automatically from the wider parts of the pores. When the soil is re-wetted the wider parts restrict water uptake, filling only at low suctions; when filled, the water will pass into the narrower parts. As a result the soil holds more water at the same suction during drying than during wetting. This is known as hysteresis.

As illustrated in Figure 2.5, water rises higher in a narrow tube than in a wide one. When the height, h, and the diameter, d, of the tube are in metres,

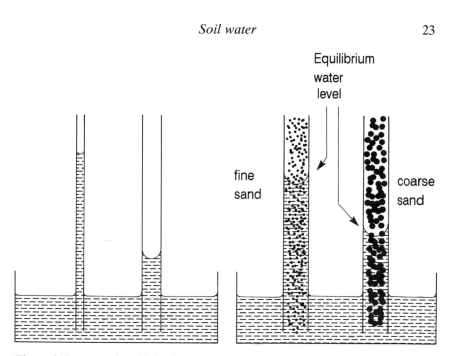

Figure 2.5 Water rises higher in a narrow capillary tube than in a wide one, and higher in granular material with small pores (fine sand) than with large pores (coarse sand).

$$h = 3 \times 10^{-5}/d. \qquad (2.2)$$

As an example, for a cylindrical tube of diameter 3 μm, h equals 10 m. The pressure, or suction, required to empty the tubes is inversely proportional to d. Similarly in soils, for a given total porosity and applied suction, the amount of water retained by soils is greater the narrower the pores. Water in the soil pores is held with a suction that depends on the pore diameter, and only when the applied suction exceeds this can water be removed. This suction is often expressed in bars, kilopascals, or metres of water (1 bar = 10 m water = 100 kPa); see Table 2.2. In a pore with a capillary rise of 10 m, a suction of 10 m would be required to empty it, and the water in the pore is held with a maximum suction of 10 m or 1 bar.

Water potential

For some purposes, and especially for vertical movement of water, it is convenient to consider the energy level of water in soil rather than the suction (negative pressure) within it. The concept is that of water potential, which requires explanation.

Soil components

Table 2.2. *Units of soil water suction and its equivalents as matric potential and cylindrical pore diameter*

Suction (m)	Suction (bar)	Matric potential (kPa)	Equivalent pore diameter
10^{-2}	10^{-3}	-10^{-1}	3 mm
10^{-1}	10^{-2}	-1	300 μm
1	10^{-1}	-10	30 μm
10	1	-10^2	3 μm
10^2	10	-10^3	300 nm
10^3	10^2	-10^4	30 nm
10^4	10^3	-10^5	—
10^5	10^4	-10^6	—

The forces acting on water in soil change its energy, that is, its ability to do work. The difference in energy between free water and water in soil, when expressed per unit quantity of water, is called potential. There are three forces that act on soil water and contribute to its total potential. These are: capillary forces, which give a capillary (more usually called a matric) potential; gravity, which gives gravitational potential; and osmosis, which gives osmotic potential. These various potentials are additive to give total water potential. The usefulness of potential is that it is a unifying concept, bringing together the various forces that make water move.

It may help in understanding matric (capillary) potential and its ability to do work, by considering a tube from which water is removed (Fig. 2.6). As the meniscus retreats from a to b over a distance l, the volume of water removed is $\pi r^2 l$. The amount of work done to remove the water is $2\pi r F l$ (remembering that work = force × distance) where F is the attractive force per unit length of contact between water and the

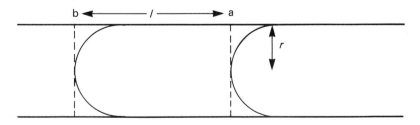

Figure 2.6 Schematic representation of water in a tube. The work required to move the meniscus from a to b is equal to $2F/r$ (see text).

wall of the tube. It follows that the amount of work done in extracting unit volume of water is given by $2\pi rFl/\pi r^2 l = 2F/r$. Hence, if the pore radius is reduced ten-fold, the amount of work required to extract unit volume of water is increased by a factor of 10. More generally, the narrower the pores containing unit volume of water, the greater the amount of work required (per unit volume) to remove it.

When expressed as energy per unit volume of water, potential has the same dimensions as suction or pressure so that potentials are often expressed in pressure units. But since work has to be done against capillary forces to extract water from soil and so return it to its free state, matric potentials are negative, and 1 bar suction is equivalent to -1 bar matric potential. When concerned with water movement, potentials are usually expressed as the height of an equivalent water column (Section 6.2). Under many conditions the matric potential is the main component of total water potential and is often regarded as the equivalent of suction.

Available water capacity

In describing the amount of soil water available to plants two terms are commonly used. *Field capacity* is the water content of soil after it has been saturated in the field, then covered to prevent evaporation and allowed to drain for 48 hours. The *permanent wilting point* is the water content of soil at which leaves of plants growing in the soil do not recover from wilting when placed in a saturated atmosphere. At field capacity the matric potential is about -5 kPa (-0.05 bar), and the permanent wilting point corresponds to a matric potential of about -1.5×10^3 kPa (-15 bar). Although neither term has an exact physical meaning, each has practical value in marking the limits of soil water available to plants. The water that can be held between the two limits is known as the *available water capacity* of the soil.

The water in soil may move to plant roots or be evaporated from the soil surface; after a heavy rain or application of irrigation it may pass through the soil. The rate at which the water moves is discussed in Chapter 6.

2.5 The soil solution

The liquid phase in soil is not pure water. It contains solutes and dissolved gases and for some purposes is better referred to as the soil

Table 2.3. *Values at 25 °C of the constant, K_H, in Henry's Law for the solubility of gases in water*

Gas	K_H (mol m^{-3} atm^{-1})	Gas	K_H (mol m^{-3} atm^{-1})
CO_2	34.1	NO	1.88
CH_4	1.50	O_2	1.26
NH_3	5.76×10^4	SO_2	1.24×10^3
N_2O	25.6	H_2S	1.02×10^2

Note: Example: CO_2 concentration in the atmosphere is 350 ppmv = 350×10^{-6} atm. From Equation 2.3, the equilibrium concentration of CO_2 in pure water is $K_H P_G = 34.1 \times 350 \times 10^{-6} = 1.19 \times 10^{-2}$ mol m^{-3}, or 0.52 mg per litre.
Source: From Stumm, W. and Morgan, J.J. 1981, *Aquatic Chemistry*. Wiley, New York.

solution, for example in considering the transfer of nutrients to plants and the transfer of elements to ground water or the atmosphere. Some of the factors that determine its concentration, which varies with time and between soils, will now be discussed.

When there is equilibrium between the soil air and soil water the distribution of a gas can be described by Henry's Law:

$$K_H = [G_S]/P_G, \tag{2.3}$$

where G_S is the concentration of gas in solution, P_G is the partial pressure of the gas in soil air and K_H is the equilibrium constant. Table 2.3 gives values of K_H for gases which are transferred between soils and the atmosphere. Ammonia and sulphur dioxide are very soluble, hydrogen sulphide, carbon dioxide and nitrous oxide are moderately soluble, and nitric oxide, methane and oxygen have low solubility. These solubilities are referred to in later chapters.

The concentration of solutes in soils at field capacity is usually in the range 1–20 moles per cubic metre of solution in non-saline soils and may reach 50–100 mol m^{-3} in saline soils. The composition varies over a wide range, as shown in Table 2.4. There are differences in composition between soils due to differences in parent material and the conditions of formation of the soil, to the addition of fertilizers, manures, lime and irrigation water, and to deposition of salts and gases from the atmosphere. The composition also depends on the water content of the soil at the time the solution is extracted.

In cultivated soils of pH above 5 the ions usually present in greatest

Table 2.4. *Concentrations of nutrient ions commonly found in soil solutions*

	Concentration			Concentration	
Ion	(mg l^{-1})	M	Ion	(mg l^{-1})	M
Ca	10–200	2.5×10^{-4}–5×10^{-3}	NO_3–N	5–200	3.5×10^{-4}–14×10^{-3}
Mg	5–100	2×10^{-4}–4×10^{-3}	SO_4–S	10–100	3×10^{-4}–3×10^{-3}
K	1–40	2.5×10^{-5}–1×10^{-3}	H_2PO_4–P	0.01–0.60	3×10^{-7}–2×10^{-5}

concentration are Ca^{2+}, Mg^{2+}, K^+, Na^+, NO_3^-, Cl^-, SO_4^{2-} and HCO_3^-. The soil solution also contains $Si(OH)_4$; NH_4^+ is sometimes present, and in acid soils it often contains Al^{3+}, $AlOH^{2+}$ and Mn^{2+}. Under anaerobic conditions Mn^{2+} and Fe^{2+} are usually present. Micronutrient ions (see Section 7.5) are present in small concentrations and the solution may also contain organic acids and dissolved organic matter.

Some of the ions in solution associate with other ions to form a soluble complex. Two examples are:

$$Ca^{2+} + SO_4^{2-} = CaSO_4; \tag{2.4}$$

$$Al^{3+} + OH^- = Al(OH)^{2+}. \tag{2.5}$$

Complexes are also formed between cations and organic acids. Formation of soluble complexes increases the solution concentration of nutrient metals such as iron and can thus improve the supply to plants.

The composition of soil solutions is buffered by clay and organic matter. This is discussed in Chapter 4.

2.6 Soil air

As mentioned above, the fractional air-filled pore space, e_a, is equal to $(e-\theta)$ where e and θ are the total fractional pore space and the volumetric water content respectively. For plants that require well-aerated soil, e_a and θ should be about equal at field capacity, air occupying the larger spaces.

Soil air has a higher concentration of carbon dioxide and a lower concentration of oxygen than the atmosphere above the soil (Table 2.5). The differences arise from respiration by soil organisms and plant roots and tend to be reduced by diffusion between the atmosphere and soil air.

Table 2.5. *Composition of the air in a poorly drained silty loam*

Sample at soil depth (cm)	Percentage composition of extracted air			
	Winter		Summer	
	CO_2	O_2	CO_2	O_2
30	1.2	19.4	2.0	19.8
61	2.4	11.6	3.1	19.1
91	6.6	3.5	5.2	17.5
122	9.6	0.7	9.1	14.5
152	10.4	2.4	11.7	12.4

Note: Note the high CO_2 concentration (by volume) and its increase with depth, also the low O_2 concentration at depth, especially in winter when diffusion from the atmosphere would be restricted by the high soil water content.
Source: From Boynton, D. and Reuther, W. 1938. *Soil Science Society of America Proceedings* **3**, 37–42.

For the complete oxidation of a carbohydrate the equation can be written:

$$[C(H_2O)]_n + nO_2 \rightarrow nCO_2 + nH_2O. \qquad (2.6)$$

The respiratory quotient, RQ, is the ratio of moles (or volume) of CO_2 produced to moles (or volume) or O_2 consumed. In this reaction, which occurs in well-aerated soils, the RQ is 1.

Diffusion of oxygen and carbon dioxide occurs because of the differences in concentration between the atmosphere and soil air, and a dynamic equilibrium is reached (Section 6.7). Unless the soil is drier than the permanent wilting point the soil air is saturated with water vapour.

The volume of air in soil is largely determined by the soil water content. The greater the water content, the smaller the volume of soil air and the more quickly is the oxygen depleted by respiration. Most plants are harmed if their roots are held in an oxygen-free (anoxic or anaerobic) environment. The activities of some microorganisms are also influenced by the supply of oxygen. Under anaerobic conditions they produce gases such as methane and nitrous oxide that contribute to global warming (Chapter 11), and other products of reduction including Fe^{2+}, Mn^{2+}, S^{2-}, ethylene and organic acids, some of which can injure plant roots. These processes of reduction under poorly aerated conditions are discussed in Chapter 5.

Table 2.6. *Mineral components in soil*

Primary minerals:	those from the parent material that are resistant to weathering; present mainly in the sand and silt fractions.
Secondary minerals:	the products of chemical weathering present in the clay fraction; they include aluminosilicates and hydrated oxides of iron, aluminium and manganese, collectively known as clay minerals.

Note: The term 'clay minerals' is often used to refer to the aluminosilicates alone as these dominate the clay fraction in most soils.

2.7 Mineral components

The minerals present in the sand and silt fractions are predominantly determined by the mineralogical composition of the parent material and the degree of weathering. The most common mineral in these fractions is quartz (SiO_2), but in soils that have not been strongly weathered micas and feldspars may also be present. Minerals such as ilmenite, zircon and haematite are resistant to weathering and can be found in soils that have been strongly weathered and leached.

The main effects of the sand and silt fractions are on the physical properties of soils. In soil dominated by coarse sand, water retention is small and drainage is rapid. At the other extreme of a soil dominated by silt, pore spaces are small so water transmission is slow, root growth may be restricted and the soil may become waterlogged after rain. Generally, the minerals in the sand and silt fractions have little effect on the chemical properties of soils although micas and feldspars, when present, slowly release plant nutrients such as calcium, magnesium and potassium on weathering.

The clay fraction is different. The minerals in the sand and silt fractions are the residues from the disintegration of parent material and hence are often known as primary minerals. Those in the clay fraction are the products of chemical weathering (Section 3.2); these are known as secondary minerals, or clay minerals, and consist of aluminosilicates and hydrated oxides (Table 2.6). The minerals in the clay fraction impart chemical and physical properties to soil which strongly influence its behaviour, for example in adsorbing cations, anions and pesticides and acting as a source of plant nutrients. Their properties can best be understood from a knowledge of their structure.

The aluminosilicates in the clay fraction range from crystalline, through poorly crystalline to amorphous, as shown by X-ray diffraction. All consist of repeating units of (i) a silicon atom surrounded by oxygen atoms in the form of a tetrahedron, and (ii) an aluminium, magnesium or iron atom surrounded by oxygen atoms and hydroxyl groups in the form of an octahedron. The repeating units are linked to form sheets called tetrahedral and octahedral sheets, which are chemically combined.

In the poorly crystalline and amorphous aluminosilicates, imogolite and allophane respectively, these units are combined with little discernible order but have an extensive surface, characteristically of hydroxyl groups.

There are three types of crystalline aluminosilicates in the clay fraction, each with layers stacked one above the other, resulting in platy crystals, often with extensive planar surface but little edge surface. In the three types each layer consists of the following:

1 : 1 clay minerals, 1 tetrahedral : 1 octahedral sheet, as in kaolinite and halloysite;
2 : 1 clay minerals, 2 tetrahedral : 1 octahedral sheet, as in pyrophyllite, illite, vermiculite, and smectites, e.g. montmorillonite;
2 : 2 clay minerals, 2 tetrahedral: 1 octahedral sheet plus an octahedral interlayer of magnesium hydroxide or aluminium hydroxide, as in chlorite.

The sequence of atoms in each layer is shown in Figure 2.7. This figure also shows the substitution of one element for another in 2 : 1 minerals. In the tetrahedral sheet silicon may be replaced by aluminium, and in the octahedral sheet aluminium may be replaced by magnesium and iron. The pyrophyllite structure is electrically neutral, but substitutions make the structure electrically negative because the substituting ion has a lower positive charge. Substitution of tetrahedral Si^{4+} by Al^{3+}, or octahedral Al^{3+} by Mg^{2+}, gives a single negative charge which is balanced by a cation held between the silicate layers. The substitution is described as isomorphous because the substituting ion must be of similar size.

Isomorphous substitution is not significant in 1 : 1 minerals, and only partial in 2 : 1 and 2 : 2 clay minerals. The negative charge that results from substitution is 'permanent', in that it is independent of the soil pH. Depending on the pH, however, all aluminosilicates in the clay fraction also dissociate protons from broken and edge surfaces, leaving the struc-

					K$^+$—	
O$_6$		O$_6$		O$_6$		
Si$_4$		Si$_4$		(Si$_3$Al)		tetrahedral
O$_4$(OH)$_2$		O$_4$(OH)$_2$		O$_4$(OH)$_2$		
Al$_4$		Al$_4$		(Al, Mg, Fe)$_4$		octahedral
(OH)$_6$		O$_4$(OH)$_2$		O$_4$(OH)$_2$		
—	—	Si$_4$		(Si$_3$Al)		tetrahedral
		O$_6$		O$_6$		
		—	—	—	K$^+$—	

(a)	(b)	(c)

Figure 2.7 Atoms in the unit cell of (*a*) kaolinite, (*b*) pyrophyllite, and (*c*) a 2 : 1 mineral showing isomorphous substitution. The unit cell is repeated in two dimensions to form sheets: the sequence of atoms would be that 'seen' looking at the sheets edge on, the sheets being stacked on top of each other. In (*c*), potassium ions balance the negative charge due to isomorphous substitution. Note: pyrophyllite is rarely found in soils but is included to show the composition of a 2 : 1 mineral with no isomorphous substitution.

ture negatively charged. They therefore have pH-dependent negative charge although in 2 : 1 and 2 : 2 clay minerals the permanent charge is much greater. The total negative charge is responsible for the cation exchange properties of clays, which are discussed in Chapter 4.

One other property of the 2 : 1 clay minerals needs to be mentioned. The extent of isomorphous substitution is illite > vermiculite > smectite. Illite has a high charge and the alumina/silica layers are held tightly together by potassium ions. With smectites the layers are more weakly held, water molecules can enter the interlayer space and cause the mineral to swell. This has implications for the cation exchange properties of the minerals. The property of swelling when wet and shrinking when dry also causes movement of the foundations of buildings, roads and bridges built on soils with high contents of smectites.

Also constituents of the clay fraction are the hydrated oxides of Fe, Al and Mn, which are often important in highly weathered soils. Commonly occurring minerals include goethite, $FeO(OH)$; haematite, Fe_2O_3; gibbsite, $Al(OH)_3$; and birnessite, a manganese oxide of variable composition. Generally they occur in poorly crystalline form. Depending on the pH they can be either positively or negatively charged (Section 4:1).

2.8 Organic components

Soil usually contains several million bacteria per gram, as well as many fungi and other microorganisms, and bigger organisms such as insects and earthworms. They play a vital role in the cycling of nutrient elements and in maintaining soil fertility. The organisms themselves and the processes they bring about are discussed in Chapter 5. The discussion here is about the non-living organic components.

In the top 10 cm, soil usually contains about 1–3% carbon in organic compounds; there is often more in soils under grass or trees, and up to about 30% in peats. Carbon compounds are added to soil in the products of photosynthesis in roots, stems, leaves, woody material, and in animal residues including faeces and sewage sludges. Some of these materials are decomposed only slowly and can be seen by eye or with a hand lens. Because of its low specific gravity (about 1 g cm^{-3}) this visible material is sometimes called the light fraction (Figure 2.8) to distinguish it from humus, most of which is adsorbed by clay and which together have a higher specific gravity. The end product of decomposition is mainly carbon dioxide which returns to the atmosphere.

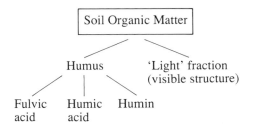

Figure 2.8 The fractions conventionally characterized in soil organic matter.

As a result of decomposition by the soil organisms and other chemical processes discussed in Chapter 5, the anatomical structure of the plant materials is lost and some of the carbon compounds are changed to the comparatively stable product called humus. This is a dark-coloured material, which shows no ordered structure when examined by X-ray diffraction. From its physical properties it is deduced that it consists of loosely interwoven strands of high-molecular-mass substances.

Determining the chemical structure of humus has proved difficult and

Table 2.7. *Average chemical composition of soil humic substances*

Component	Humic acid	Fulvic acid
C(%)	56	46
O(%)	36	45
H(%)	4.7	5.4
N(%)	3.2	2.1
S(%)	0.8	1.9
COOH (mmol g^{-1})	3.6	8.2
Phenolic OH (mmol g^{-1})	3.9	3.0

Source: From Schnitzer, M. and Khan, S.U. (eds) 1978, *Soil Organic Matter*. Elsevier, Amsterdam; with permission.

more research is needed before a satisfactory account can be given. Four main lines of work can be identified.

1. Humus can be divided into fractions that differ in solubility in acids and alkalis. Most of the humus can be extracted with alkaline solutions. The precipitate that forms on acidification of the extract to pH 2 is known as humic acid, and the fraction remaining in solution is known as fulvic acid. Another fraction, humin, is not extracted with alkali. These fractions are not single, chemically distinct substances, but they differ between themselves in elemental composition and in the amounts of reactive groups they contain (Table 2.7).

2. Various chemical compounds and groups of compounds have been identified in humus. For example, about 10–15% of humus carbon is present in polysaccharides (biopolymers consisting of sugar units). Some are residual from plants, but the identity of the sugar components indicates that most are synthesized by the soil microorganisms. The sugar units contain sugar acids known as uronic acids, which appear to act as a 'glue' in cementing soil particles together in aggregates.

3. Partial oxidation, reduction and acid hydrolysis yield a bewildering array of products, including aromatic and aliphatic substances. Although there is always the possibility that some of the products were formed during the extraction and chemical treatments, the use of methods that preserve the original structure of humus has confirmed the presence of aromatic and aliphatic structures. Amino acids are also products of hydrolysis and account for about one half of the nitrogen in humus. Some of the sulphur in humus is also present in amino acids. Organic phosphorus compounds present in humus include inositol phosphates and smaller amounts of nucleic acids and phospholipids.

4. The humic and fulvic fractions of humus have acid groups of which the carboxylic (COOH) and phenolic (OH) groups are present in largest amounts. The carboxyl groups dissociate between about pH 4.5 and 7, and the phenolic OH groups at higher pH. Thus for carboxyl groups:

$$-COOH \rightleftharpoons -COO^- + H^+. \tag{2.7}$$

Humus therefore has a negative charge, which is pH-dependent and provides a source of cation exchange properties.

Summary of the properties of humus

1. Because of the presence of $-COOH$ and $-OH$ groups, which dissociate, humus has pH-dependent negative charges. At pH 7 in well-drained soils the value is mainly due to the dissociation of carboxyl groups and is usually about 3 moles of charge per kilogram of humus. It therefore contributes to the retention of cations and some pesticides as well as to the pH buffering properties of soils (Chapter 4 and Section 9.3).

2. The plant nutrient elements nitrogen, phosphorus and sulphur are contained in organic compounds. In the top 10–15 cm of soils the ratio of % organic carbon : % N, the C/N ratio, is usually between 10 and 14, the C : organic P ratio is about 100, but differs between soils, and the C : organic S ratio is usually about 80–100. When the organic compounds are mineralized by the action of microorganisms these three elements are released as inorganic ions, which may then be taken up by plants.

3. It reacts with metal cations to form complexes, some of which are soluble and some insoluble. The stability of these complexes varies with pH but at pH 5 is reported as:

$$Cu > Pb > Fe^{2+} > Ni > Mn > Co > Ca > Zn > Mg.$$

Trivalent metals (Fe^{3+} and Al^{3+}) are held strongly.

4. It interacts with clay minerals and oxides of iron and aluminium to form stable aggregates, thereby improving the soil physical conditions for plant growth.

Although higher plants can grow to maturity in the absence of any external organic substances, these properties of humus make it a key component in the successful management of soils.

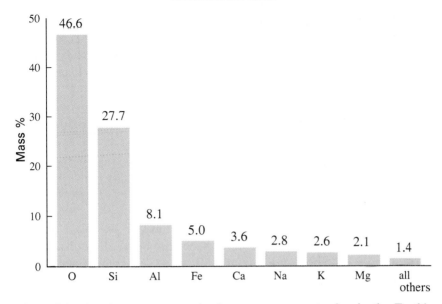

Figure 2.9 The elements present in the greatest concentration in the Earth's crust. (Data from Mason, B. and Moore, C.B. 1982. *Principles of Geochemistry*, 4th edition. Wiley, Chichester).

2.9 Chemical elements in soil

The elements present in the greatest concentration in the Earth's crust are shown in Figure 2.9. In soil, the order of occurrence of the 12 elements present in greatest concentration is, *on average*,

O> Si > Al > Fe = C = Ca > K > Na > Mg > Ti > N > S.

All other elements, including those that are essential plant nutrients, are present in concentrations of less than 0.1%. Whereas the concentration in soil of potassium is about 1.5%, that of the micronutrient molybdenum is only about 0.00001%.

As would be expected from the minerals present in soils (Section 2.7), the most abundant elements are those that form the aluminosilicates. The only pronounced difference in average composition between soils and the rocks that form the Earth's crust is the much greater concentration of C, N and S in soils as a result of biological accumulation. In individual soils there can, however, be big differences in composition between the parent material (rock) and the derived soil.

2.10 Interactions between soil components

In order to understand soil properties, the components have been described individually. One component, however, often influences the properties of others and also the processes that occur in soils. To give two examples, soils with high contents of coarse sand retain only small amounts of water, and the good aeration results in rapid oxidation of added organic matter to carbon dioxide; in such soils the content of organic matter is therefore often low. Secondly, 1 : 1 clay minerals have a smaller capacity to retain acidic inputs than 2 : 1 clay minerals; if the organic matter contents are the same, soil with 1 : 1 minerals will develop a low pH more quickly.

The important processes that occur in soils are discussed in later chapters and reference is made throughout the book to the interaction between components and processes. First, we discuss soil formation and account for the wide range of soil properties that are found under field conditions.

2.11 Summary

The fabric of the non-living part of soil is made up of mineral and organic materials, and contains pore space occupied by water and air. The minerals include residues from the parent material, usually mainly quartz, and minerals in the clay fraction, which are the products of weathering. These clay minerals give the soil important properties including the capacity to adsorb cations and other solutes on negatively charged surfaces. Organic matter has this same property and is also a source of the plant nutrients nitrogen, phosphorus and sulphur.

The physical properties of soil depend largely on the sizes of the soil particles (soil texture) and on their arrangement (soil structure), as in aggregates. To a large extent texture and structure determine the distribution and movement of water and air in soil, and the availability of water to plants. They also affect the growth of plant roots.

The soil water contains solutes and dissolved gases, and may therefore be referred to as the soil solution. The solutes it contains include nutrients, which are taken up by plant roots.

3

Development of soils

3.1 Introduction

Most of the soils at temperate latitudes in the northern hemisphere have been formed during the past 10 000 years. Pre-existing soils were stripped away by glaciers or eroded by rain, melting snow and fast-flowing rivers during the Ice Ages when the land supported no more than sparse vegetation. Only a few remnants of the soils from the earlier warm interglacial periods can be found.

At southern latitudes where there was no extensive glaciation, as in Australia, soils bear clear evidence of changes of climate. Below the present soil profile can sometimes be seen an earlier profile from which the top soil was lost by erosion and on which wind-borne material was deposited.

New soils are being formed and others are being eroded at the present day; for example, eroded soil from the Himalayas is carried down rivers to form alluvial soils in Bangladesh. Fine sand from the Sahara Desert is blown into Europe, and soil is formed from recently deposited volcanic ash and recent flows of lava. Some of the most fertile soils are formed on recently deposited alluvium, as in the Nile Valley, which is derived from soil and rocks in the Ethiopian highlands. These examples, and others could be given, show that soils have a transient existence on a geological time scale.

This chapter deals with the formation of soils; erosion of soil is dealt with in Chapter 12. The processes of soil formation are usually separated into (i) rock weathering, and (ii) the addition of organic matter and the formation of structures that characterize soils, as will be done here. The two processes do, however, take place concurrently, as may soil erosion. First to be dealt with is the formation of the soil mineral components.

3.2 Rocks and rock weathering

Three broad groups of rocks are generally recognized: igneous, sedimentary and metamorphic. Igneous rocks are formed by the cooling of the earth's liquid magma. A variety of igneous rocks is formed according to the chemical composition of the magma and its rate of cooling. Granites, for example, are formed by slow cooling of magma high in silica; they contain large proportions of quartz and feldspars, which are present as large crystals. In contrast, basalts form from magma low in silica; they contain minerals such as amphiboles, pyroxenes and olivines and the crystal size is small because the magma cooled rapidly. Table 3.1 gives the general chemical formulae of these minerals; it should be noted that the actual composition of some can differ, though within strict limits, according to the composition of the magma from which they crystallized.

Sedimentary rocks, as the name implies, were laid down as sediments.

Table 3.1. *The main minerals in igneous rocks and the primary minerals in soils*

Name	Chemical formula	Presence in soil
Quartz	SiO_2	Most common mineral in sand and silt fractions
Orthoclase feldspar	$KAlSi_3O_8$	Present in weakly and moderately weathered soils
Muscovite (a mica)	$K(Si_3Al)Al_2O_{10}(OH)_2$	
Biotite (a mica)	$K(Si_3Al)(Mg, Fe)_3O_{10}(OH)_2$	
Pyroxenes	$(Mg, Fe)SiO_3$	Easily weathered to form clay minerals
Amphiboles	$(Mg, Fe)_7(Si_4O_{11})_2(OH)_2$	
Olivines	$(Mg, Fe)_2SiO_4$	

The material is the product of rock weathering and erosion, and has therefore gone through at least one weathering cycle. As a result of particle segregation during transport, sedimentary rocks usually have a narrow range of particle size and they can therefore be described as sandstones, siltstones, or clays. Sedimentary rocks can also be formed by chemical precipitation and deposition of the remains of animals and plants, chalk and limestone being the most commonly occurring examples.

Metamorphic rocks are the third group recognized by geologists. High temperatures and pressures cause some degree of recrystallization of sedimentary rocks, converting sandstones into quartzite and clays into slate. The geological formation called Basement Complex, which underlies much of the land mass of Africa, consists of partly metamorphosed igneous rocks.

Soils are formed from the three groups of rocks, and also from deposits known as drift or till that are left by glaciers, from loess, alluvium in valley bottoms and colluvium on the lower slopes of hills.

Weathering processes

Physical weathering causes rocks and minerals to disintegrate. The processes, depending on the climate to which the rock is exposed, include differential thermal expansion and contraction of the different minerals, expansion of cracks when water freezes, and abrasion by glaciers and wind-borne sand. Disintegration increases the surface area of the rock and its minerals, allowing chemical decomposition to proceed more quickly.

Several processes can be involved in chemical decompositions, as shown by the following examples.

(*a*) *Hydrolysis* of orthoclase feldspar:

$$4KAlSi_3O_8 + 4H^+ + 18H_2O \rightarrow Si_4Al_4O_{10}(OH)_8 + 4K^+ + 8Si(OH)_4. \quad (3.1)$$
orthoclase　　　　　　　　　　　kaolinite　　　　　　　silicic acid

Where there is through-drainage, potassium ions and silicic acid are removed and kaolinite is left as a residue. Other clay minerals can be formed, as discussed below.

(*b*) *Oxidation* of ferrous iron (Fe^{2+}) to ferric iron (Fe^{3+}):

$$Fe^{2+} \rightarrow Fe^{3+} + e^- \text{ (electron)}. \quad (3.2)$$

Biotite, a member of the mica group of minerals, contains Fe^{2+} as a constituent of its mineral lattice. During weathering the Fe^{2+} is oxidized and forms a coating of hydrous ferric oxide, $Fe(OH)_3$, on the surfaces of the weathering products. (See also Section 5.8.)

(*c*) *Hydration*, for example of quartz:

$$SiO_2 + 2H_2O \rightarrow Si(OH)_4. \tag{3.3}$$

(*d*) *Carbonation*, as in sedimentary rocks containing calcium carbonate:

$$H_2O + CO_2 \rightleftharpoons H_2CO_3; \tag{3.4}$$

$$CaCO_3 + H_2CO_3 \rightleftharpoons Ca(HCO_3)_2. \tag{3.5}$$

In areas of chalk and limestone the calcium bicarbonate is dissolved in leaching water (it is one component of 'hard' water). The calcium carbonate is reprecipitated if the temperature of the solution rises, the partial pressure of CO_2 in the atmosphere decreases, or the solution evaporates; hence the formation of stalagmites and stalactites. Increased atmospheric concentrations of CO_2, as has been occurring for the last few decades, will increase the rate of carbonation.

Decomposition is also induced by lichens (associations of fungi and algae), soil microorganisms and the roots of higher plants. The carbon dioxide released by respiration into the soil air dissolves in water and lowers the pH, increasing the hydrolysis of minerals. The production of organic acids by these organisms has the same effect. Organic acids and humus also form stable complexes with many metals, which assist in their removal from the mineral structure. The effect of these biological agents is therefore to increase the rate of decomposition, especially in soils with strong biological activity.

3.3 The products of weathering

All the chemical reactions referred to above depend on the presence of water. In a completely dry environment there is no decomposition of rock minerals and the weathering products are rock and mineral fragments which have undergone no chemical change.

At a given temperature, the longer the weathering rock remains wet the greater is the degree of chemical weathering. The effect of temperature itself is mainly on the rate of weathering, although temperature also affects the nature of the products that are formed. Where the weathering zone is wet for much of the year and temperature is high, as in the humid tropics, there is a deep layer of weathered rock and mineral fragments between the bottom of the soil profile and the rock. This weathered material, saprolite, is readily eroded by water and only accumulates at sites that are physically stable. Under these conditions depths of 10 m and more of saprolite have been observed.

Minerals in the clay fraction are formed during weathering, by:

(i)   modification of the structure of primary silicates (an example is the formation from micas of illite, also known as hydrous mica, by the replacement of potassium by ions that are more hydrated);

(ii) synthesis from the products of weathering (kaolinite may be formed by this process in addition to its formation by hydrolysis as shown in Equation 3.1);

(iii) extreme weathering, which leaves only the least soluble components such as haematite, Fe_2O_3, and gibbsite, $Al(OH)_3$.

Smectites (Section 2.7) form where calcium, magnesium, silica and aluminium are present together, usually because weathering has taken place under semi-arid conditions, or the products have accumulated in low-lying areas. In upland areas of the wet tropics the most soluble products of weathering are leached, and kaolinite is formed where there is sufficient silica. Two non-crystalline clay minerals – allophane and imogolite – are formed from volcanic ash and, in time, form kaolinite and its hydrated form halloysite.

In the comparatively young soils of northern Europe, northern Asia and North America, soil clay fractions usually contain illite, mixed-layer silicates and goethite (a hydrated iron oxide). Under more intense weathering, kaolinite is dominant where there is free drainage, and oxides of iron, aluminium and titanium dominate where weathering is extreme.

To summarize: the effects of physical and chemical weathering are three-fold.

1. Solid rocks become fragmented, providing niches for plants, which add organic matter to the mineral material.
2. The mineral composition changes owing to (*a*) the loss of readily decomposed minerals, including plagioclase feldspar, the ferromagnesian minerals such as hornblende and augite, and biotite mica, and (*b*) the formation of clay minerals. The mineral composition of soils therefore depends on the minerals in the parent material, the severity of weathering and whether the soluble products are leached or accumulate.

3. The chemical composition is also changed. Comparison between the composition of rocks and their weathered products at individual sites, show that sodium, calcium and magnesium and to a lesser extent potassium, are leached in through-drainage.

The factors that determine the composition of the residual inorganic components are parent material, climate (temperature and rainfall), time (the duration of weathering), topography (e.g. upland compared with low-lying areas) and biological factors. This theme is returned to in Section 3.6.

3.4 Additions and decomposition of organic matter

If no further organic matter were added, decomposition would eventually reduce the organic matter content of soil to zero, although it would take several hundred years to oxidize all the stabilized humus. In practice, organic matter is added from the above-ground parts of plants and trees and also from roots (Section 5.2). The content of organic carbon in soil is therefore set by the quantity of organic matter that enters the soil each year and the rate of decomposition of the soil organic matter.

Examples of accumulation and decrease of organic carbon are shown in Figure 3.1. The annual change may be expressed as:

$$dC/dt = A - rC, \tag{3.6}$$

where A is the annual addition of organic carbon, r is the decomposition constant, that is, the fraction of organically held carbon decomposing to CO_2 each year, and C is the amount of organic carbon already in the soil.

When the rates of addition and loss are equal, $dC/dt = 0$ and $A = rC_e$, where C_e is the content of organic carbon at equilibrium (Figure 3.1a). This provides a means of calculating r as A/C_e. If, for example, the input (A) of carbon to an old arable field under wheat is 1.7 t ha^{-1} a^{-1} and it contains 34 t of organic carbon to a depth of 23 cm (C_e), then $r = 1.7/34 = 0.05$. In other words, 5% of the organic carbon decomposes each year.

The dynamics of soil organic matter are described by the half-life ($t_{1/2}$) and turnover time (t_e) of soil carbon:

$$t_{1/2} = 0.693/r; \tag{3.7}$$

$$\text{and } t_e = C_e/A = 1/r. \tag{3.8}$$

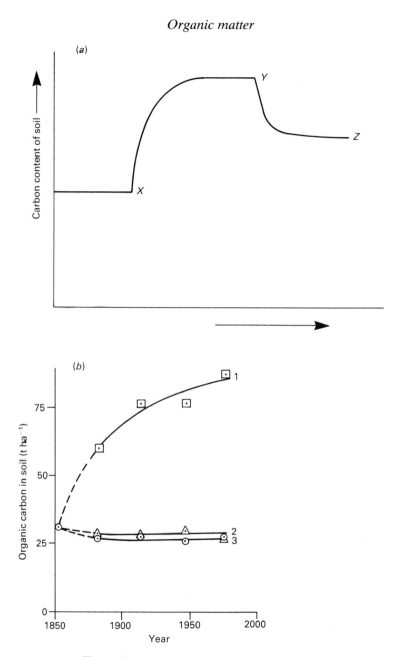

Figure 3.1 The organic matter content of soils.
(a) The equilibrium content, X, is changed to Y, a new equilibrium value, and later to Z, as the cultivation system is changed (schematic).
(b) Effects of annual addition of farmyard manure (1), and fertilizers (2) on the organic carbon content of soil compared with that in unmanured soil (3). Data from Hoosfield, Rothamsted Experimental Station.

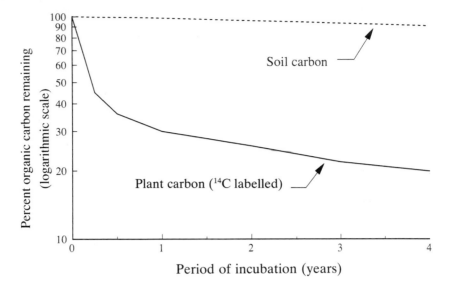

Figure 3.2 Comparison of rates of decomposition of soil organic matter and cut grass incubated in soil. (Modified from Jenkinson, D.S. 1965. *Journal of Soil Science* **16,** 104–15.)

The half-life is the length of time needed for a given input of organic carbon to fall to half its initial level. The turnover time, which is the average age of the soil carbon, is defined as the amount of carbon in soil at equilibrium divided by the annual input. In the example given, the *average* age of the soil carbon is $1/0.05 = 20$ years.

It is well known that plant residues freshly added to soil are decomposed faster than soil humus. Figure 3.2 shows the comparative rates of decomposition of ryegrass and humus in the same soil, the ryegrass being labelled with ^{14}C (a radionuclide; see Section 10.7) to allow its loss to be distinguished from the loss of unlabelled soil $C(^{12}C)$. Results like those in Figure 3.2 can be used to establish values of r for successive stages in the decomposition process and then to improve the calculation of the rate of change of soil carbon content.

3.5 Processes of soil formation

Soils begin to form as soon as plants establish themselves on a mineral substrate. This might be the fragmented rock developed *in situ* by weathering, debris deposited by ice, wind or rivers, or land reclaimed

Table 3.2. *Processes of soil formation*

1. Weathering of parent material
2. Addition and partial decomposition of organic matter
3. Formation of structural units
4. Leaching and acidification
5. Clay eluviation
6. Podzolization
7. Desilication
8. Reduction
9. Salinization and alkalization
10. Erosion and deposition of eroded soil

Note: Processes 1, 2 and 3 occur in the formation of all soils; the occurrence of processes 4–10 depends on the environmental conditions.

from the sea. If the conditions for plant growth are suitable the content of organic matter can increase quickly, as shown in Figure 3.1*a*. Plant leaves and stems fall on the surface of the soil, and a high density of roots develops near the soil surface. Normally therefore, the content of organic matter is highest near the soil surface and decreases with depth.

It was mentioned in Section 2.8 that humus and clay interact to form aggregates. Channels are created by the larger members of the soil fauna, e.g. by earthworms burrowing through the soil, and channels are left when plant roots decay. Cracks are formed when clay soils containing smectites dry. The arrangement of the mineral particles (soil structure) is therefore different from that in weathered mineral debris. Several processes bring about these and other changes (Table 3.2). The processes that dominate at a site depend on the environmental conditions; some will now be referred to.

(i) *Leaching and acidification*

If there is excess rain so that water drains through the soil profile, soluble substances are removed. Also, even in an unpolluted atmosphere, rain water is acidic because carbon dioxide dissolves in it to form carbonic acid. Several processes in addition to acidic rain lead to soil acidification, as discussed more fully in Section 9.5.

In acid soils fungi become dominant members of the microbial population. The composition of the soil fauna is also changed; for example, burrowing earthworms become fewer in number if the pH is below

5. One result is to slow down the decomposition of added organic matter. It may accumulate at the surface in a form known as mor humus in which the structure of plant residues is strongly retained. It is distinguished from mull humus, formed under near-neutral conditions, which is more humified and more evenly distributed through the soil profile.

(ii) *Clay eluviation*

In many parts of the world clay is washed from the upper part of the soil profile and deposited lower down. The upper, clay-depleted part is known as the A or eluvial horizon (eluvial means 'washed from') and the lower part is the B or illuvial horizon (illuvial means 'washed into'). The clay is transported by water moving through conducting pores. It is deposited in the B horizon, usually in an ordered fashion as clay skins on the walls of pores. Deposition might be caused by a decrease in the rate of flow of water or by changes in chemical conditions making the suspension of clay unstable. Argillic (clay) horizons can also form by weathering processes, but these do not give the ordered feature of clay skins. Argillic horizons occur commonly in the soils of the world; they are present in Alfisols and Ultisols (Section 3.7).

(iii) *Podzolization*

'Podzol' is derived from the Russian, often translated as 'below ash'; the ash refers to the bleached, ash-grey colour of the upper, A, horizon (Figure 3.3). The process of podzolization occurs in acid soils. Low-molecular-mass organic acids, fulvic acid, and polyphenols from plant leaves form soluble complexes with iron, aluminium and clay. The organic and inorganic components are transported by water and deposited in sharply defined B horizons. Other processes have been proposed but this has most experimental support.

Podzolization occurs on freely drained sites under coniferous forest, beech and oak trees (and several other tree species), and also under heath plants such as *Calluna* and *Erica* species. It occurs typically in cool, humid regions. The profile is readily recognized in the field and the process has been studied widely. The term 'podzol' is still in use but is now largely replaced by the almost equivalent term 'Spodosol'.

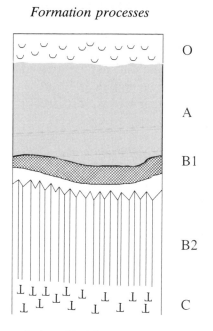

Figure 3.3 Diagram of a podzol profile. The sequence of horizons is: O (organic layer); A (light-coloured eluvial horizon); B1 (illuvial horizon which might consist of a layer with a high content of organic matter overlying a layer with a high content of iron oxide); B2 (horizon stained by tonguing with organic matter and iron oxide); C (horizon little affected by the podzolization process).

(iv) *Desilication*

This term refers to the greater leaching of silica relative to iron and aluminium oxides. The process occurs in the tropics; it creates highly porous soils with high contents of iron oxide, and less commonly of gibbsite, $Al(OH)_3$. Oxisols are a common type of soil with these properties. They occur on old land surfaces, as in the South American cerrado.

(v) *Reduction*

Where drainage is impeded the soil air is replaced by water, the activity of the soil microorganisms becomes limited by the supply of oxygen and acids are produced by fermentation. Organic matter accumulates because its oxidation is slow under these conditions, and peat may be formed. Microorganisms reduce Fe^{3+} to Fe^{2+} so that the soil loses the brown colour of ferric oxides and becomes grey or bluish grey. The soil is described as gleyed. If reducing conditions are temporary there is only partial reduction, and the soil has a mottled appearance with brown iron oxide staining and reduction in grey areas. Chemical reduction by microorganisms has other effects, discussed in Chapter 5.

(vi) *Salinization and alkalization*

Salinization refers to the accumulation of salts such as sulphates and chlorides, and alkalization to the accumulation of sodium on cation exchange sites. Salts, blown inland from oceans, introduced in irrigation water, or produced by weathering, accumulate in depressions and may render the soil infertile. They are a particular problem of irrigation in hot countries if insufficient water is used to leach salts from the profile. If the salinity is largely due to sodium salts, alkalization occurs when Ca^{2+} and Mg^{2+} are either precipitated as carbonates or leached out, and Na^+ occupies 10–15% or more of the cation exchange sites. Under these conditions the soil structure collapses, pH rises above 8 and cultivation becomes very difficult (see also Section 8.10).

(vii) *Erosion and deposition*

As has been mentioned earlier, soils have only a transient existence on a geological time scale. They are always liable to be eroded through the action of wind or rain (Chapter 12). The eroded material might be deposited in river valleys to provide the starting material for new soil, or it might be lost into the oceans. Wind-blown material might be similarly lost or be deposited on top of existing soil. It follows that processes (i)–(vi) above do not proceed smoothly for ever. Further, they are often interrupted by a change of climate, interference by man and for other reasons. Considered next, therefore, are the factors that affect soil formation.

3.6 Soil-forming factors

In outlining the soil-forming processes in the previous section, reference was made to the influence of the parent material of the soil, rainfall, temperature, and other conditions in the environment. All these conditions are factors that determine a soil property. The factors were put into an equation by Jenny:

$$s = f(cl, o, r, p, t, ...)$$

using s for a soil property, cl for climate, o for organisms, r for relief, p for parent material and t for time. Thus soil is a function, f, of five soil-forming factors, with more being added if needed. He later modified the equation to:

$$s = f(L_o, P_x, t)$$

where L_o is the state of the system when soil formation starts, and may include mineral and organic materials and relief; P_x is a composite of external factors and includes climate and organisms; and t is time.

The equations express complexity in a simple form. They are most useful when relating a soil property to a single soil forming factor, the others remaining constant. For example, Jenny related the nitrogen content of soils to climate. This might be done for other soil properties, but there is usually the difficulty that factors other than the one under investigation are rarely constant. The concept of soil-forming factors is nevertheless useful because it helps to describe the pattern of soil distribution in the field.

Parent material

The effect of parent material is most pronounced in young soils. It becomes less the longer the soil-forming processes operate and the more intensive they are. Soils formed on granites are coarsely sandy because of the quartz present in the granite; they are likely to have low pH and low base status and to tend towards podzolization. Basic igneous rocks contain ferromagnesian minerals which weather to clay minerals; they contain little or no quartz, and they form soils with clay or loam texture and relatively high base status and pH. Similarly, soils formed on glacial till and loess in Europe, North America and northern Asia, and soils formed on sedimentary rocks, have properties that are usually strongly influenced by the nature of the parent material.

Surface relief

On slopes, soil material is displaced downhill, where it accumulates as colluvium, leaving shallow soils on the upper slopes. Soils in depressions and valleys are usually wetter than those on slopes because either the water table is higher in the profile or water has accumulated from upslope, or for both reasons. On the terrain between valleys (inter-fluves) drainage is often impeded on hill tops which are flat or have gentle slopes. In both situations reduction may occur (see Section 5.8). Where climate and parent material are constant or vary in the same way with altitude, the pattern of soils follows the topography and is known as a toposequence.

Climate

The broad pattern of soil distribution due to climate was first noted in Russia. Over this large continental region, temperature increases from north to south and annual rainfall increases from east to west. Vegetation changes with climate but parent material and relief are relatively constant and soil properties can be related to climate. The relationship has been developed further. Soils that fitted into the climatic zones became known as Zonal soils, those where the effect of parent material dominated were called Intrazonal, and those where soil formation was restricted were called Azonal. These ideas on the factors determining the distribution of soils passed from USSR to the USA and other countries, where they were modified to suit local conditions.

Organisms

The term includes vegetation, animals and man. The effect of vegetation can be seen in temperate regions under old woodland. The soil profile usually has an organic layer of partly decomposed litter and, depending on the tree species, often shows podzolization, as described above. Under grassland the organic matter is usually mull humus and is well distributed in the profile. In temperate regions earthworms have a pronounced effect in mixing the soil components in soils that are near neutral. In the tropics termites collect plant material, and by creating mounds and underground chambers they mix the soil horizons and create large channels, which can conduct rainwater. Microorganisms are largely responsible for the decomposition of plant residues in soil, but have several other effects (Chapter 5). Finally, man is a powerful influence, radically changing soils through cultivation, drainage and irrigation, by adding lime and other chemicals, and probably most seriously by accelerating erosion.

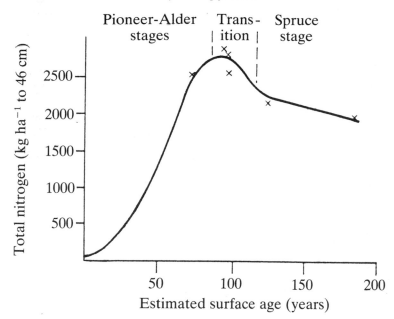

Figure 3.4 Rapid accumulation of nitrogen in soil and litter layer after the recession of ice from moraines in southeast Alaska. (From Crocker, R.L. and Major, J., 1955. *Journal of Ecology* **43**, 427.)

Time

This is obviously a factor because most soil properties depend on the period of time over which the soil processes have acted. The clock may start with the retreat of a glacier, a fall of volcanic ash, the burning of forest trees, or from other events which can be dated. An example is shown in Figure 3.4. The rate of soil formation is, however, more complex than it might first appear. It depends on the soil property under investigation; thus, the organic matter content changes more rapidly than the assemblage of clay minerals. It also depends on the parent material, slope, and the other soil forming factors. It is referred to again in Section 12.4 in relation to the 'acceptable limit' of soil loss by erosion.

The soil-forming factors and soil processes provide means of generalizing about soils. Their limitations are: (i) they are usually not sufficiently well characterized at a particular site; and (ii) the conditions of formation will almost certainly have changed during the soil's history. An alternative is to use soil properties, as discussed in the next section.

3.7 Soil classification

The purpose of soil classification is to group soils with similar properties so that the name of the group is useful in conveying information about its members. It is much the same as in classifying plants and animals: communication is only possible if we know what is meant by, say, domestic cat, even though cats come in different shapes, sizes and colours. If a name is put to a soil group it should convey information that applies to all the members of the group, even though the members may differ in other respects.

Many systems have been devised to classify soils. Each country has tended to develop its own system in order to assess its soil resources, but for international understanding more general systems are needed. The two most widely used have been devised by FAO (put forward as a legend to the FAO/UNESCO *Soil Map of the World*, but really a classification scheme) and in the USA by the US Department of Agriculture. One or other has now been adopted by several countries.

The USDA system was described in 1975 in *Soil Taxonomy*. The highest level of classification is the Order, of which there are ten (Table 3.3). Below the Order are successively Suborder, Great Group, Subgroup, Family and Series. Most of the ten Orders are characterized by diagnostic horizons. For example, Spodosols (previously known as podzols) are characterized by the spodic horizon in which there is an illuvial accumulation of free iron and aluminium oxides and organic matter. At lower levels of classification, distinction is made according to the soil moisture and temperature regime, base status, degree of expression of horizons and other properties.

The FAO system has 26 classes which are roughly equivalent to Suborders and Great Groups in *Soil Taxonomy*. Cross-referencing between the two systems is therefore often possible. FAO retained earlier names such as Podzol and Chernozem and introduced new ones.

A big difficulty in establishing a system of classification which is useful throughout the world is that many areas have not been surveyed. Unless there has been a survey on the ground, soil properties have to be inferred from aerial photographs and satellite data. These techniques provide a way to extrapolate from ground-based observations but are not a substitute for them.

Table 3.3. *Land area of the Soil Orders used in* Soil Taxonomy

Soil Order[a]	Some of the properties of each Order	Mha	%
Alfisols	Soils with clay translocation and deposition (high base status)	1730	13.1
Aridisols	Soils in dry climates where salts and carbonates may have accumulated	2480	18.8
Entisols	Recently formed soils with limited development of horizons	1090	8.2
Histosols	Peat and fen soils	120	0.9
Inceptisols	Young soils with few diagnostic features	1170	8.9
Mollisols	Temperate grassland soils	1130	8.6
Oxisols	Highly weathered soils of the tropics	1120	8.5
Spodosols	Soils with subsoil accumulation of humus and iron/aluminium oxides	560	4.3
Ultisols	Soils with clay translocation and deposition (low base status)	730	5.6
Vertisols	Swelling clay soils	230	1.8
Total		10 360	
Mountains		2810	
Total		13 170	

[a]An additional Soil Order of Andisols is being proposed for inclusion. For more complete information see Buol, S.W., Hole, F.D. and McCracken, R.J. 1989. *Soil Genesis and Classification*, 3rd edition. Iowa State University Press, Ames.

3.8 Summary

Soil material consists of minerals resistant to weathering, clay minerals formed by weathering, and organic matter. It may also include calcium carbonate, calcium sulphate and other chemical compounds.

The composition of soils and the properties of soil profiles depend on the nature of the parent material, the processes that have taken place since soil formation started, and the relative rates of addition and decomposition of organic matter.

A relationship has been described relating a soil property to five soil-forming factors; for example, soil nitrogen content has been related to rainfall and temperature. Often there is insufficient information about conditions of soil formation (and erosion) in the past to establish this type of relationship.

There are large differences between soil profiles from place to place over the surface of the Earth. To describe soil profiles in a coherent manner, various classification schemes have been introduced.

4

Sorptive properties of soils

As a result of weathering and the addition of organic debris, soils contain clay minerals and organic matter. These two components adsorb ions, molecules and gases. Important consequences of this property are that soils act as a buffer zone between the atmosphere and groundwater, and provide plants with a steady supply of nutrients. Soils have this sorptive property because of the electrical charges and large surface area of the clay minerals and humus, as will first be described.

4.1 Electrically charged surfaces

Because of their structure and chemical composition, the clay minerals and humus usually bear a negative charge (Chapter 2). As shown in Table 4.1, the charge can be permanent, that is, it remains the same whatever the solution pH as long as the structure remains intact. Only the 2 : 1 clay minerals have a predominantly permanent charge.

From the surfaces and edges of clay minerals (aluminosilicates and hydrous (hydrated) oxides of iron and aluminium) and from acidic groups in humus, protons dissociate to an extent that depends on the

Table 4.1. *Sources of electrical charge on soil clays and humus*

1. Permanent negative charge on 2:1 and 2:2 clay minerals due to isomorphous substitution.

2. pH-dependent charges:

    (*a*) negative charges at broken surfaces and edges of clay minerals

$$
\begin{array}{c}
| \\
-SiOH \\
| \\
-AlOH \\
| \\
-SiOH \\
| \\
-AlOH \\
|
\end{array}
\quad
\begin{array}{c}
OH^- \\
\rightleftharpoons \\
H^+
\end{array}
\quad
\begin{array}{c}
| \\
-Si\text{-}OH \\
| \\
-AlO^- \\
| \\
-Si\text{-}OH \\
| \\
-AlO^- \\
|
\end{array}
\quad + H_2O
$$

    (*b*) negative charges on humus due to dissociation of carboxyl and phenolic hydroxyl groups

$$-COOH \quad \underset{H^+}{\overset{OH^-}{\rightleftharpoons}} \quad -COO^- + H_2O$$

$$>COH \quad \underset{H^+}{\overset{OH^-}{\rightleftharpoons}} \quad >CO^- + H_2O$$

    (*c*) negative and positive charges on hydrous oxides of iron and aluminium

$$>FeOH \quad \underset{H^+}{\overset{OH^-}{\rightleftharpoons}} \quad >FeO^- + H_2O$$

$$>FeOH_2^+ \quad \underset{H^+}{\overset{OH^-}{\rightleftharpoons}} \quad >FeOH + H_2O$$

pH. This gives a pH-dependent negative charge, which increases as the pH is raised. Hydrous oxides of iron and aluminium can become positively charged at low pH by adsorption of protons (Table 4.1). The relationship between charge and pH for three clay minerals is shown in Figure 4.1.

Because the pH-dependent charge arises at the surfaces and edges of clay minerals, it is greater the more disordered is their structure. Kaolinite forms relatively large crystals with less surface area and fewer broken edges than allophane, which has a much more disordered structure. The charge on both minerals is pH-dependent and at the same pH

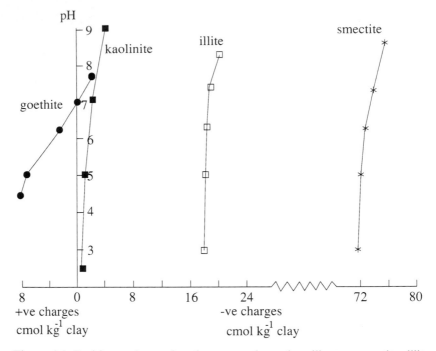

Figure 4.1 Positive and negative charges on three clay silicates: smectite, illite and kaolinite, and on goethite. (From Greenland, D.J. and Mott, C.J.B., 1978. In *The Chemistry of Soil Constituents* (eds. D.J. Greenland and M.H.B. Hayes). Wiley, Chichester.)

allophane has a much higher charge per unit mass. The surface properties of the reactive soil components are listed in Table 4.2.

Cations that balance the charge

The negative charge is exactly balanced by adsorbed cations. Some of the cations in 2 : 1 clays are held in interlayer spaces and can diffuse only very slowly into the outer solution. These cations are called 'non-exchangeable'. Figure 4.2 shows non-exchangeable potassium ions held in the narrow interlayer spaces between the silicate sheets of illite. Cations held at the edges and on the surfaces of the particles have ready access to the outer solution and are described as 'exchangeable'.

The 2 : 1 clay minerals, such as illite, which hold non-exchangeable cations have a high charge density, that is, a high charge per unit surface area, and the silicate sheets are held tightly together. Smectites have a lower charge density and so the silicate sheets are not held together so

Table 4.2. *Surface properties including cation exchange capacity (CEC) at pH 7 of the reactive soil components*

	Approximate surface area $(m^2\ kg^{-1})$	Approximate CEC $(mol_c\ kg^{-1})^a$	Approximate surface charge density $(\mu mol_c\ m^{-2})$	Dominant negative charge
Kaolinite	$(1-2)\times10^4$	0.02–0.06	1–6	pH-dependent
Illite	1×10^5	0.3	3	permanent
Smectites	8×10^5	1.0	1	permanent
Vermiculite	8×10^5	1.4	2	permanent
Fe and Al oxides	3×10^4	0.005	0.2	pH-dependent
Allophane	$(5-7)\times10^5$	0.8	1.5	pH-dependent
Humic acid	9×10^5	3.0	3	pH-dependent

[a]Units for CEC are $mol_c\ kg^{-1}$ (moles of charge per kilogram); they are sometimes expressed as $cmol_c\ kg^{-1}$, that is, centimoles (0.01 mole) of charge per kilogram. In the older literature the units are usually milliequivalents (meq) per 100 g soil; 1 meq per 100 g = 1 $cmol_c\ kg^{-1}$. Note that for monovalent ions, e.g. Na^+, 1 mole = 1 equivalent; for divalent ions, e.g. Ca^{2+}, 1 mole = 2 equivalents; for trivalent ions, e.g. Al^{3+}, 1 mole = 3 equivalents.

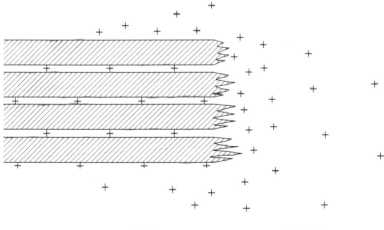

Illite   Solution

Figure 4.2 Exchangeable and non-exchangeable cations in clays. The cations (+) held near the outer surfaces of clay particles are exchangeable; some clay minerals, e.g. illite, contain potassium ions in the interlayer spaces, which are non-exchangeable.

strongly. As a result, each interlayer space is wider and cations in this space can readily diffuse into the outer solution. The important consequence is that smectites, with more available surface area, hold more exchangeable cations than illite, which has a higher charge density.

4.2 Exchangeable cations and cation exchange capacity

As described in the previous section, the negative charge of humus and part of the negative charge of clay minerals varies with pH. The capacity of soils to hold exchangeable cations therefore also depends on the pH and is usually measured in the laboratory in one of two ways.

1. The cation exchange capacity (CEC) is measured at a standard pH, usually 7. This is most conveniently done by passing a solution of ammonium acetate, which is buffered at pH 7, through the soil to replace all the exchangeable cations by ammonium ions. The amounts of displaced cations can be determined by standard chemical methods and the total is the cation exchange capacity. Alternatively, the surplus ammonium acetate is washed out of the soil and then the exchangeable ammonium is displaced by another cation and measured.

2. The CEC is measured at the soil pH by displacing the exchangeable cations with a solution of potassium chloride, which is unbuffered. All the displaced cations are determined and their summation is the cation exchange capacity, now known as the effective cation exchange capacity (ECEC).

Method 1 may be regarded as giving a standard result because it is independent of the variability of soil pH. Method 2, however, gives a value more related to actual soil conditions. It has been found to be useful in characterizing soils in which most of the negative charge is pH-dependent, as with many soils in the tropics.

The clay fraction and humus each contribute to the cation exchange capacity of a soil. Because of interaction of their surfaces it is not strictly correct to calculate the CEC of a soil from the summation of the CECs of the components. The summation is nevertheless useful in showing the relative importance of the contributions of organic matter and clay minerals to CEC, and the effect of pH. Two examples are:

1. In a soil of pH 7 with 10% clay composed of illite, which has 0.3 mole of charge per kg ($0.3 \text{ mol}_c \text{ kg}^{-1}$) and 3% humus ($3.0 \text{ mol}_c \text{ kg}^{-1}$) the CEC by addition is $(0.3 \times 0.1) + (3.0 \times 0.03) = 0.03 + 0.09 = 0.12 \text{ mol}_c \text{ kg}^{-1}$, the contribution of the humus being greater than that of the clay.

If the humus content falls to 1% the CEC of the soil becomes 0.06 mol_c kg^{-1}.

2. In a soil of pH 4 with 10% kaolinite clay (*ca.* 0.04 mol_c kg^{-1}) and 3% humus, the CEC is close to zero because the negative charges on both components are pH-dependent and tend towards zero at pH 4. The result is the same if the clay minerals are iron and aluminium oxides. In many soils of the tropics these oxides or kaolinite dominate the clay fraction so that when they are acid the soils have very low cation exchange capacities. With the illitic clay, as in example 1, the CEC is largely pH-independent so that at pH 4 the CEC of the soil would be about 0.03 mol_c kg^{-1}. This is one reason why soil acidification can be more serious in the tropics than in temperate regions where soils usually contain 2 : 1 clay minerals, which have largely permanent negative charge.

The exchangeable cations present in a sample of soil can be determined by the displacement methods referred to above that are used to measure CEC. The amounts vary greatly between soils; amounts in a soil from Senegal (Table 4.3) are lower than in many soils

Table 4.3. *Exchangeable cations in a soil profile from Senegal*

Depth (cm)	Exchangeable cations				Bulk density (g cm^{-3})
	Ca	Mg	K	Na	
(a) Exchangeable cations in cmol kg^{-1}					
0–6	1.13	0.53	0.15	0.05	1.3
6–13	0.85	0.33	0.10	0.05	1.3
13–31	0.85	0.05	0.05	trace	1.4
31–79	1.05	0.03	0.05	trace	1.4
(b) Exchangeable cations in kg ha^{-1}[a]					
0–6	353	99	46	9	
6–13	309	72	35	10	
13–31	857	30	49	—	
31–79	2822	48	131	—	
Total	4341	249	261	19	

[a]Example of calculation: for Ca, 1.13 cmol kg^{-1}
$$= 1.13 \times 0.40 \text{ g kg}^{-1} = 1.13 \times 0.40 \times 10^{-3}\text{g g}^{-1} \text{ soil.}$$
$$= 1.13 \times 0.40 \times 10^{-3} \times 1.3 \text{ g cm}^{-3} \text{ soil.}$$
To 6 cm depth the volume of soil per hectare is 6×10^8 cm^3.
Amount of Ca is therefore $1.13 \times 0.40 \times 10^{-3} \times 1.3 \times 6 \times 10^8 \times 10^{-3} = 353$ kg ha^{-1} to 6 cm depth.

of temperate regions, but down to rooting depth they are considerably in excess of the requirements of an annual crop, at least for calcium and magnesium.

4.3 The diffuse layer

In Figure 4.2 the cations held in solution outside the clay structure help to balance the negative charge but are positioned only schematically. The spatial distribution of the exchangeable cations can be described in terms of two opposing forces: (i) attraction toward the charged surface, which raises the cation concentration close to the surface; and (ii) diffusion away from the surface towards the outer solution because of the difference in concentration. The result is that, in the presence of sufficient water, many cations including Na^+, Mg^{2+} and Ca^{2+}, and to some extent K^+ and NH_4^+ are distributed in a diffuse layer, as shown in Figure 4.3. The ions in this layer and those in the outer solution are in dynamic equilibrium. Because Cl^- and NO_3^- are repelled from negatively charged surfaces their concentrations are lower in the diffuse layer than in the outer solution.

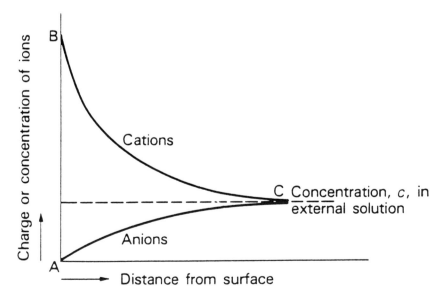

Figure 4.3 Representation of the diffuse layer of cations and anions close to a clay surface. Note that the concentration of cations increases, and the concentration of anions decreases, close to the surface.

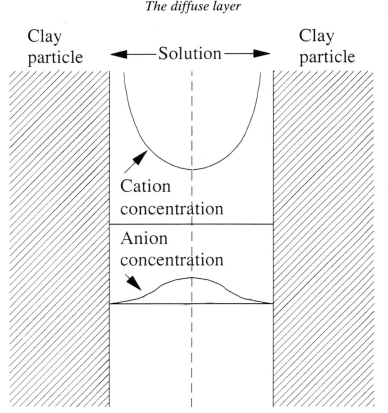

Figure 4.4 The concentrations of cations and anions between two clay particles close together in aqueous suspension.

The diffuse layer becomes compressed in solutions of high concentration, mainly because diffusion away from the surface is less. The properties of the cation also have an effect: the higher its charge the more strongly is it attracted to the surface, and the greater its hydration the more weakly is it attracted. Sodium ions are weakly held because of their single charge, and in solution each has a large shell of water molecules. Calcium ions are more strongly held and so the diffuse layer of a calcium-saturated clay is more compact than that of a sodium-saturated clay. One implication of the theory of the diffuse layer will be described.

Consider the distribution of sodium ions in a suspension of 2 : 1 sodium-saturated clay (Figure 4.4). When the surfaces of the clay particles come close together the diffuse layers of sodium ions overlap; this overlap increases their concentration. There is then a tendency for water to be imbibed to restore the original distribution of ions. As a consequence the clay particles repel one another. If the diffuse layer is greatly compressed, as in a salt solution of high concentration, or by replacing Na^+ by Ca^{2+}, the clay particles can approach each other closely and then stick together. This is one of the processes that occurs in the formation of soil aggregates.

Aggregation of clay particles does not occur in a soil with a high proportion of exchangeable sodium ions (above about 15% of the CEC) and a low salt content because the particles repel one another, as described above. In this kind of soil, known as sodic, the soil particles are dispersed, giving adverse physical conditions (Secton 8.10).

The theory of the diffuse layer does not apply to cations and anions that form a complex with the mineral (or humus) surface itself. Their reactions are discussed in Sections 4.5 and 4.6.

4.4 Characteristics of cation exchange

Experiments in the middle of the nineteenth century showed that when a dilute solution of ammonium chloride was added to the top of a column of soil, a dilute solution of calcium chloride came out of the bottom. The observation was important because it showed that ammonium sulphate, then being introduced as a nitrogen fertilizer, would not be wasted by being quickly washed out of the soil. The reaction, assuming a soil saturated with calcium, and using sodium chloride instead of ammonium chloride, may be represented by the equation:

$$Ca \text{ soil} + 2\ NaCl \rightleftharpoons Na_2 \text{ soil} + CaCl_2. \qquad (4.1)$$

The characteristics of this and other cation exchange reactions are:

(i)   it is rapid;
(ii)   it is reversible;
(iii) exchange is of equal cation charge (mole/valency);
(iv)   the distribution of two species of cation, e.g. Na^+ and Ca^{2+}, between soil and solution depends on their relative concentrations and force of attraction to the negatively charged surfaces of clay and humus.

Equation 4.1 can be used to describe the equilibrium between two cations, as when a solution of a sodium salt is shaken with calcium-saturated soil. A common form of the equation, using X to represent soil, reducing the numbers by half, and using the subscript (aq) to represent ions in solution, is

$$Ca_{1/2}X + Na^+_{(aq)} \rightleftharpoons NaX + \tfrac{1}{2}\ Ca^{2+}_{(aq)}. \tag{4.2}$$

At equilibrium,

$$K^G = \frac{\surd[Ca^{2+}_{(aq)}]}{[Na^+_{(aq)}]} \cdot \frac{[NaX]}{[Ca_{1/2}X]}, \tag{4.3}$$

where the square brackets indicate moles of charge per unit mass of soil and moles of charge per unit volume of solution; K^G is called the *Gapon constant*. It is only constant over a limited range of cation ratios. Each soil has its own value of K^G for each pair of cations. Equation 4.3 is commonly used to calculate the effect of saline irrigation water on the amount of sodium adsorbed onto soil.

It follows from Equation 4.3 that if the ratio $[Na^+_{aq}]/\surd[Ca_{aq}]$ is kept constant, the composition of the soil expressed as $[NaX]/[Ca_{\frac{1}{2}}X]$ will also be constant, at least in the range where K^G is constant. One example of the use of Equation 4.3 is as follows:

For exchange between H^+ and Ca^{2+} the equation is

$$K^G = \frac{\surd[Ca_{(aq)}]}{[H^+_{(aq)}]} \cdot \frac{[HX]}{[Ca_{1/2}X]},$$

which may be written as $\dfrac{\surd[Ca_{(aq)}]}{[H^+_{(aq)}]} = K^G \cdot \dfrac{[Ca_{1/2}X]}{[HX]}.$ \hfill (4.4)

Expressed in logarithmic form, the ratio $\surd[Ca_{(aq)}]/[H^+_{aq}]$ becomes $(pH - \tfrac{1}{2}\, pCa)$, known as the lime potential. The concentration of protons (H^+) and hence the pH of a soil solution will be seen to vary with the concentrations of other cations in solution. To overcome this problem the pH can be measured on a suspension of soil in a salt solution that keeps the cation concentration about constant; in soils that are about neutral this is done by measuring the pH in 0.01 M calcium chloride. It is, however, more convenient, and the common practice, to measure the pH of a suspension of soil in water.

Equation 4.3 can be expressed as follows: 'When cations in solution are in equilibrium with a larger number of exchangeable cations, a change in the concentration of the solution will not disturb the equilibrium if the concentrations of all the monovalent ions are changed in one ratio, those of the divalent ions in the square of that ratio and those of all the trivalent ions in the cube of that ratio'. This is usually known as the Ratio Law. It can be applied to the soil solution which becomes more concentrated when a soil dries. For K^G to remain constant, divalent and trivalent ions pass into solution and monovalent ions become adsorbed. Similarly, if the soil becomes wetter monovalent ions pass into solution and divalent and trivalent cations become adsorbed. The relative concentrations of cations, in addition to the overall concentration of the soil solution, therefore varies as the soil becomes wetter or drier.

The adsorption of trace amount of cations has only a small effect on the amounts of other cations which are adsorbed. If this effect is small enough to neglect, adsorption can be shown as a plot of amount of adsorbed ion, x, against the equilibrium solution concentration, c. The slope of the line (x/c) is known as the adsorption coefficient, K. The relation $K = x/c$, or $x = Kc$, is a simple form of an equation known as the Freundlich equation. This type of plot is used for the adsorption of organic molecules, as referred to in discussing the adsorption of pesticides (Section 13.4).

4.5 Selectivity of cation adsorption

The affinity of most cations for an adsorbing surface is greater (i) for divalent than for monovalent ions, and (ii) for large cations than for small ones of the same charge because the larger the cation the less hydrated it is. The usual affinity series is:

$$Al^{3+} > Ba^{2+} > Sr^{2+} > Ca^{2+} > Mg^{2+} = Cs^+ > Rb^+ > K^+ = NH_4^+ > Na^+.$$

The meaning of selectivity is illustrated in Figure 4.5.

For K^+ and NH_4^+ the type of clay mineral affects their selectivity and ease of replacement. Their size (radii of K^+ and NH_4^+ are 0.13 and 0.14 nm, respectively) allows them to fit into cavities in the tetrahedral layers of 2 : 1 minerals, where they are strongly held. If the clay dries they are effectively trapped and only slowly diffuse to the outer solution.

Metals such as copper, zinc, chromium, manganese, iron, cobalt and nickel form complex ions in solution, for example MOH^+ (where M is the metal). These metals form similar complexes at surfaces that contain hydroxyl groups, especially those of hydrated iron, manganese and aluminium oxides. The complex ions do not undergo cation exchange,

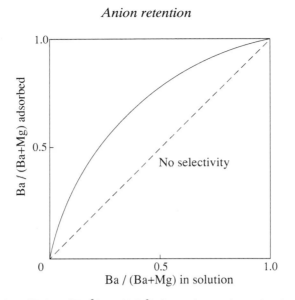

Figure 4.5 The affinity of Ba^{2+} and Mg^{2+} for a clay surface. On the surface the proportion of Ba^{2+} is greater than in solution.

as described in Section 4.4, and can only be displaced by extraction with acids or strong complexing agents.

4.6 Anion retention

Figure 4.3 shows the repulsion of anions from surfaces that have a dominantly negative charge. At low pH positive charges are developed on the surfaces of hydrated oxides of iron and aluminium (Table 4.1) and to a less extent on 1 : 1 clay minerals. Under these conditions Cl^-, NO_3^- and other anions are adsorbed and undergo exchange with each other. Adsorption is, however, often blocked by competition from organic anions.

A more important mechanism for the retention of anions is termed 'ligand exchange'. This refers to the formation of a surface complex between an anion and a metal, usually Fe or Al, in a hydrated oxide or clay mineral. For example:

$$- Fe - OH_2^+ + H_2PO_4^- \rightarrow - Fe - O - PO_3H_2^0 + H_2O;$$

$$- Fe - OH_2^+ + HPO_4^{2-} \rightarrow - Fe - O - PO_3H^- + H_2O.$$

These reactions increase the negative charge on the surface. Silicate, sulphate and probably fulvic acid are adsorbed similarly to phosphate. Sulphate forms a complex weaker than that formed by phosphate, which can therefore displace it. The phosphate complex is particularly strong so that as a consequence the concentration of phosphate ions in the soil solution is low. They can be displaced by high concentrations of other strongly held anions such as fluoride and silicate, but under field conditions the release of phosphate may depend partly on the generation of organic acids or OH^- by microorganisms.

4.7 Adsorption of organic molecules

The two reactions of importance are between pesticides and soil and between clay and humus. Those of pesticides are discussed in Section 13.4.

Humus is adsorbed onto clay surfaces by reactions which are not fully understood. One reaction that probably occurs depends on the presence of divalent or trivalent cations, which act as bridges between particles of negatively charged clay and negatively charged humus.

4.8 Sorption of gases

All gases dissolve in soil water to some extent. Their distribution between the gaseous and solution phases is described by Henry's Law (Section 2.5).

The solution of gases in soil water should be described as sorption by the soil rather than adsorption (sorption at surfaces). Ammonia is a very soluble gas and dissolves to form ammonium hydroxide. It is also adsorbed at the surfaces of clay and humus, displacing water from the hydration shells of exchangeable cations and reacting with protons on the surfaces of clay minerals and probably humus. Carbon dioxide forms carbonic acid, sulphur dioxide forms sulphurous acid and then sulphuric acid, and nitrogen dioxide forms nitric acid, all being agents in soil acidification (Chapter 9). The reactions of carbon monoxide and ozone with soil are referred to in Chapter 11. As will be referred to in later chapters, soil acts as both source and sink for gases.

4.9 Summary

Soils are able to adsorb cations, several species of anions, organic substances and gases. Because of this property soils act as a buffer between the atmosphere and ground water. Adsorbed ions replenish the soil solution when uptake by plant roots occurs.

The capacity of soils to adsorb cations depends on the negative charge of clay minerals (aluminosilicates and oxides of iron and aluminium) and humus. It also depends on the pH especially for the kaolin group of clay minerals, humus and oxides, the negative charge becoming less as the pH decreases.

In most soils used for agriculture calcium and (to a lesser extent) magnesium are the dominant exchangeable cations. If sodium becomes the dominant cation (more than about 15% of the cation exchange capacity) humus and clay particles becomes dispersed and the soil structure collapses.

5

Organisms and soil processes

5.1 Introduction

A great variety of organisms live in the soil. Soil animals, generally referred to as the soil fauna, range in size from earthworms and termites to those that can only be seen with a good hand lens. Microorganisms can only be observed under an optical or electron microscope; although small in size their numbers are large (Table 5.1 and Figure 5.1).

Most of the organisms depend on the addition of carbon compounds in plant materials (roots, leaves and stems) and the faeces of animals. By decomposing these materials carbon dioxide and mineral nutrients are produced, which plants take up through their leaves and roots respectively, some of the carbon remaining in the soil in organic matter. This cycling of carbon and mineral nutrients between soils and plants is a characteristic of terrestrial ecosystems.

Soil microorganisms are also responsible for some specific reactions to be referred to later in this chapter. Two examples are the oxidation of NH_4^+ to NO_3^-, and the fixation of nitrogen gas. One group can photosynthesize, making them independent of the addition of organic

Table 5.1. *Groups of organisms present in soils*

Note that microorganisms are millions per gram of soil and animals are millions per hectare.

Microorganisms in fertile soil[a] (millions g^{-1})		Animals in mull soil under beech		
			millions ha^{-1}	% of total animal mass[b]
Bacteria	1–100	Earthworms	1.8	75.1
Actinomycetes	0.1–1	Enchytraeid worms	5.3	1.5
Fungi	0.1–1	Gastropods	1.0	7.0
Algae	0.01–0.1	Millipedes	1.8	10.6
Protozoa	0.01–0.1	Centipedes	0.8	1.8
		Mites and springtails	44.1	0.4
		Others	7.2	3.6

[a]The carbon in soil microorganisms accounts for about 3% of the soil organic carbon, that is, they contain about 1 tonne of carbon per hectare in a soil with 1.5% carbon to 15 cm depth.
[b]Total mass of animals was 286 kg.
Source: The numbers of animals are from Brown, A.L. 1978, *Ecology of Soil Organisms.* Heinemann, London.

materials. Although the organisms that inhabit soil have diverse activities, the greatest number are involved in the decomposition of plant materials.

5.2 Organic materials: sources and decomposition

The organic materials added to soil are the products of photosynthesis by higher plants. Some of the carbon in compounds that are photosynthesized is returned to the atmosphere as CO_2 by plant respiration. The rest is known as net primary production (NPP); some of this is stored in perennial tissues such as wood, some may be eaten by herbivores, and usually a large part is shed as litter. Table 5.2 gives some values for a range of ecosystems. To take tropical forest as an example, the annual NPP has a mean value of 30 tonnes of dry matter per hectare per year (about 14 t C ha^{-1} a^{-1}) and at steady state (when there is no net annual addition to the mass of trees or understorey) the input of litter to the forest floor is also 30 t ha^{-1} a^{-1}. (Note: this equivalence between NPP and litter fall does not take account of the decay and death of roots, which can be substantial, but is difficult to measure.) In grasslands of

Table 5.2. *Net primary production (NPP), total biomass and its distribution, and litter input, for six types of vegetation*

Rows 1 and 2 show net primary production, rows 3 and 4 show the mass of vegetation, rows 5, 6 and 7 show the distribution of this mass, and row 8 shows the litter input. In this set of data net primary production is made equal to the production of litter.

	Tundra	Boreal forest	Temperate deciduous forest	Temperate grassland	Savanna	Tropical forest
1. NPP (mean) t ha^{-1} a^{-1}	1.5	7.5	11.5	7.5	9.5	30
2. NPP (range) t ha^{-1} a^{-1}	0.1–4.0	4–20	6–25	2–15	2–20	10–35
3. Biomass (mean) t ha^{-1}	10	200	350	18	45	500
4. Biomass (range) t ha^{-1}	1–30	60–400	60–600	2–50	2–150	60–800
5. Photosynthetic %	13	7	1	17	12	8
6. Wood %	12	71	74	0	60	74
7. Root %	75	22	25	83	28	18
8. Litter input (t ha^{-1} a^{-1})	1.5	7.5	11.5	7.5	9.5	30

Note: From Swift, M.J., Heal, O.W. and Anderson, J.M. 1979. *Decomposition in Terrestrial Ecosystems*. Blackwell Scientific Publications, Oxford; with permission.

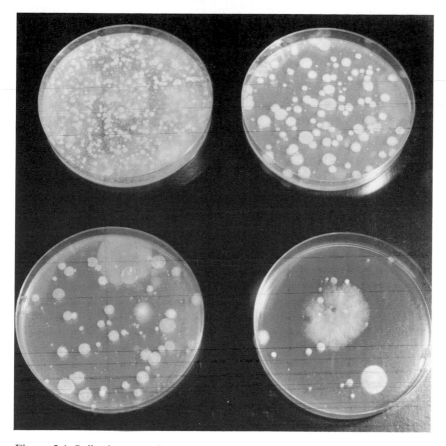

Figure 5.1 Soil microorganisms growing on agar plates. Each shows the growth 7 days after the plates were inoculated with a suspension of soil. Top left, 0.1 cm^3 suspension of 1 g soil in 100 cm^3 water added to plate (10^3 × dilution); top right 10^4 × dilution; bottom left 10^5 × dilution; bottom right 10^6 × dilution. At 10^6 × dilution 28 colonies of bacteria can be observed so that 1 g soil contained $28 × 10^6$ bacteria capable of growing on agar. (Photograph P.J. Harris.)

temperate regions the NPP and input of litter are one quarter as great. The values vary between sites; those in Table 5.2 are included only as an indication of their approximate size. During establishment and growth of the vegetation, litter production equals NPP minus increment of dry matter.

The values in Table 5.2 are for natural ecosystems. In arable agriculture the NPP is little different, but removal of the useful part of the crop for consumption, and burning of the stover where this is practised, means that the addition of organic materials to the soil is much less.

Depending on the climate of the region, the addition of dry matter from crops in roots and stubble is about 0.5 to 5 t ha^{-1} a^{-1} (0.2 to 2 t ha^{-1} a^{-1} of carbon); if the straw is returned to the soil the total addition can be 1–10 t ha^{-1} a^{-1} (0.4 to 4 t ha^{-1} a^{-1} of carbon).

As mentioned above, most of the soil organisms depend on organic compounds deposited on the soil, or in it by roots; a small amount may also be added in rain-wash from the leaf canopy. Metabolism of the organic compounds provides the organisms with energy for growth, locomotion and reproduction, and the carbon required for the synthesis of cell components. The organisms also use other nutrient elements, for example, nitrogen, phosphorus and sulphur in the plant materials in synthesizing their tissues.

5.3 Soil fauna

The soil fauna can be classified in terms of body length as micro- (less than 0.1 mm), meso- (0.1 to 10 mm) and macro- (greater than 10 mm). Protozoa are the only important members of the microfauna and are referred to in Section 5.4. Animals which burrow into soil but spend much of their time outside it are not included in the soil fauna. Examples of soil and litter fauna are shown in Figures 5.2, 5.3, 5.4 and 5.5*a*.

Earthworms, members of the Lumbricidae, are macrofauna, and are often present in large numbers in soils. In Table 5.1 the population in a soil under beech is given as 1.8 million per hectare; in productive pastures counts of about 10 million per hectare have been recorded. Their mass per hectare can reach about 2 tonnes, about the same as that of animals grazing the pasture.

Lumbricus terrestris is one of the commonly occurring earthworms. It pulls leaves into its burrows, it can migrate from near the soil surface to a depth of about 2 m to escape desiccation, and can be found in soils with a pH of about 4 though it is more abundant when the soil pH is near 7. The formation of casts on the soil surface is usually due to a few species of *Allolobophora*. The casts include partly digested plant material, and because of respiratory loss of CO_2 they have lower C : N and C : organic P ratios than the soil as a whole, thus accelerating the formation of mineral nutrients for plants. Several species create channels, which act as transmission pores for water after heavy rain or irrigation (Figure 5.5*b*). Some earthworms do not burrow but live in the litter layer or surface soil, and some inhabit compost and manure heaps.

The mesofauna are represented by very many members, the composi-

(a)

(b)

Figure 5.2 (a) A millipede, *Julus scandinavius*, a common inhabitant of leaf litter; length 20 mm.
(b) Two pill millipedes, *Glomeris marginata*, in rotting wood with faecal pellets to the right; diameter of each rolled-up millipede 5 mm.
Millipedes have bacteria in their gut which probably allow them to partly degrade lignin and cellulose. (Photographs © S. Hopkin.)

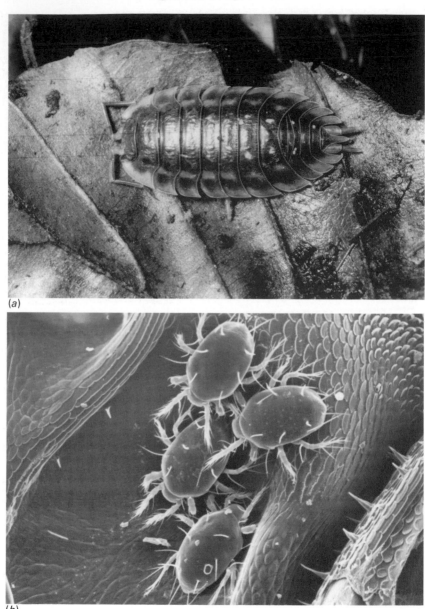

(a)

(b)

Figure 5.3 (*a*) A woodlouse, *Oniscus asellus*, an inhabitant of leaf litter; length 12 mm.
(*b*) Four mites, members of the Order Acari, riding on the underside of a woodlouse; length of each mite 0.15 mm. Mites can occur in large numbers in leaf litter and soil; various species eat plant material, plant roots and fungal hyphae. (Photographs © S. Hopkin; (*b*) by SEM.)

(a)

(b)

Figure 5.4 Springtails (Collembola) of (*a*) the family Sminthuridae, length 1 mm and (*b*) the family Entomobryoidea, length 4 mm. They occur in large numbers in leaf litter and soils; they eat fungal hyphae and may stimulate new hyphal growth. (Photographs © S. Hopkin.)

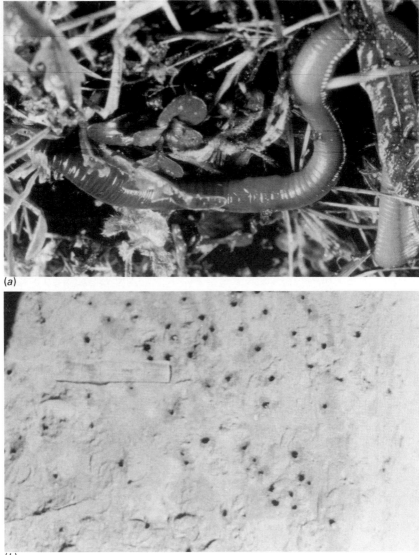

(a)

(b)

Figure 5.5 (a) An earthworm, *Lumbricus terrestris*, diameter about 7 mm, on the soil surface.

(b) Earthworm channels stained with aniline blue on the horizontal surface of a sandy loam, arable soil at a depth of 69 cm. To show the continuity of the channels the top 2 cm of soil was removed. The horizontal surface was cleaned by vacuum and the dye was poured into the channels. The horizontal surface at 69 cm was then exposed and photographed. The number of continuous channels was 238 per m² and the surface area they occupied was 0.7% of the total. Their diameter ranged between 2 and 9 mm; rule = 150 mm.

Table 5.3. *Commonly used terms that describe the physiological requirements of soil microorganisms*

Heterotrophs use organic compounds as their source of carbon.

Autotrophs use carbon dioxide or carbonates as their source of carbon. Two groups are:

Photoautotrophs, which obtain their energy from photosynthesis;

Chemoautotrophs, which obtain their energy from the oxidation of compounds.

Zymogenous organisms proliferate rapidly in the presence of an organic substrate.

Autochthonous organisms consume comparatively resistant organic materials at a steady rate.

Aerobic organisms require oxygen gas for respiration; usually called obligate aerobes.

Anaerobic organisms will grow only in the absence of oxygen; usually called obligate anaerobes.

Facultative organisms can adapt to the presence or absence of oxygen.

tion of the population depending on the environmental conditions of temperature, soil water content, aeration and pH, and the composition of the leaf litter. Collembola (springtails) and Acari (mites) are often the most abundant. The Collembola contain species that are active under dry conditions and others more active in wetter conditions, as in the woodlands of temperate zones. The total population of the meso-fauna varies greatly between sites but can exceed 2000 million per hec-tare in old grassland soils. They are active decomposers of litter, comminuting it, hydrolysing polysaccharides and depolymerizing pro-teins and lignin. Members of the population interact with each other and also with microorganisms, groups of organisms succeeding one another as decomposition proceeds.

5.4 Soil microorganisms

Five groups of microorganisms are listed in Table 5.1. Within each group there is a diversity of physiological requirements (Table 5.3). The heterotrophs require an organic substrate whereas autotrophs do not. Populations of zymogenous organisms grow rapidly when an organic substrate is added, but autochthonous organisms respond very little. Microorganisms also differ in their requirements for oxygen gas: the obligate aerobes require it for respiration, the obligate anaerobes require its absence, and the facultative organisms can adapt to its presence or absence. All the microorganisms require mineral nutrients, the list being probably the same as for higher plants (Section 7.5).

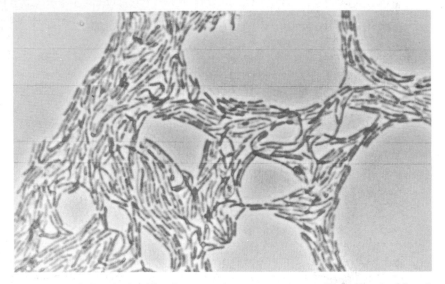

Figure 5.6 Cells of a bacillus from a colony grown on agar inoculated with soil (see Figure 5.1). The cells from the colony were spread on a glass slide; each cell is 3 μm long. (Photograph P.J. Harris.)

Bacteria

The cells of bacteria are either rod-shaped (Figure 5.6) or more nearly spherical (cocci). The rods are about 1 μm wide and up to 3 μm long, and the cocci are about 2 μm in diameter. Some bacteria change their shape with age and their habitat, which makes identification more difficult. Some form spores, which allows their populations to survive desiccation. In soils they may attach themselves to soil particles. Their distribution is, however, not uniform because populations increase where there is a source of organic substrate. Near roots, for example, the population may be 100 times higher than in the rest of the soil. This part of the soil is called the rhizosphere and its properties are discussed in Section 7.6.

The activities of the soil bacteria show great diversity. Most are heterotrophs and take part in the decomposition of plant residues. Because of their range of enzymes there are no naturally occurring organic substances, and few that are man-made, that they are unable to decompose. They are present in all soils whether these are acid or alkaline, waterlogged or well-drained and in regions that are hot or cold, wet or dry.

Actinomycetes

The cells of actinomycetes are about the same size as bacterial cells. The cells form filaments, as do fungi, but the greater similarity is with bacteria with which they are classified. All actinomycetes are heterotrophic, decomposing plant residues in soils and composts. They occur widely in soils but are restricted to an aerobic environment, and tend to be more dominant in warm soils than cold. The genus *Streptomyces* is well known as the producer of the antibiotics streptomycin and aureomycin. The actinomycete *Frankia* invades roots of alder (*Alnus*) and *Casuarina*, where it fixes nitrogen (Section 5.5).

Fungi

The cells of most fungi are joined to form filaments (hyphae) which are usually 5–20 μm in diameter and can be several centimetres or even a few metres in length. When they colonize leaf litter and compost, several hyphae often become aggregated as rhizomorphs, which can be observed by eye. More commonly observed are fruiting bodies of the group Basidiomycotina (basidiomycetes), which we know as toadstools and mushrooms.

All fungi are heterotrophs and most are saprophytic (living on dead tissues), but some invade plant roots and are plant pathogens: *Fusarium* and *Verticillium* are examples. The saprophytic fungi are active decomposers of plant residues and can decompose all components of plant material. Members of the basidiomycetes are among the few microorganisms that attack lignin, the third most abundant constituent of higher plants; they also attack cellulose and related compounds. They are known as white-rot fungi because when they attack wood they leave a light-coloured residue. Fungi are generally more tolerant of acidity than are bacteria and so tend to be the dominant decomposers in acid soils. Some fungi have a symbiotic relationship with plant roots, forming a structural association known as a mycorrhiza. The importance of mycorrhizas in plant nutrition is discussed in Section 7.7.

Algae

The cells of soil algae are about 10–40 μm in diameter; they occur singly or in colonies. All are photoautotrophs, that is, they are able to photosynthesize, and therefore tend to be concentrated near the soil surface, which may become green if left undisturbed. In most soils they are less important members of the microbial population than fungi and bacteria. They are more important as primary colonizers of mineral debris because of their photosynthetic ability, which produces the organic materials for other microorganisms. They produce polysaccharides, which help to aggregate the soil particles and hence stabilize the soil surface against erosion. This stabilization and the nutrients accumulated by the algal cells create conditions favourable for colonization by higher plants. On dry rock surfaces the algae are protected from desiccation by association with a fungus, the association being known as a lichen, which is an agent of rock weathering. Lichens are sensitive to air pollution and especially to the presence of sulphur dioxide.

Blue-green algae, now known as cyanobacteria, have the unusual property of being able to both fix atmospheric nitrogen and photosynthesize, which makes them important members of ecosystems. Like algae, they form lichens by forming an association with a fungus. They are referred to in Section 5.5.

Protozoa

Protozoa are single-celled animals, which feed selectively on bacteria and fungi; some are photoautotrophic, resembling algae, and others are saprophytic. The cells of soil protozoa are generally less than 50 μm in diameter. They are mobile and move in soil pores where there is sufficient thickness of water. Under dry conditions they form cysts, which are resistant to desiccation. Their main function in soil ecology appears to be in the control of the size of bacterial and fungal populations. There is evidence that they liberate nutrients by feeding on bacteria, an indication of the complexity of nutrient turnover in soils.

As a final comment: the microorganisms that inhabit soil, their interactions, and their effects on higher plants have not been thoroughly investigated. Much remains to be done. In the sections that follow some of the specific processes induced by microorganisms will be discussed.

5.5 Biological nitrogen fixation

Nitrogen gas is converted by a few genera of microorganisms into nitrogen compounds that can be used by plants. This process of biological nitrogen fixation provides natural ecosystems with their main source of nitrogen and is also important in agriculture. All of the organisms that

Table 5.4. *Biological nitrogen fixation by microorganisms*

	Main genera containing nitrogen-fixing species
1. Non-symbiotic	
Heterotrophic, aerobic	*Azotobacter, Azotococcus, Beijerinckia*
Heterotrophic, anaerobic	*Clostridium, Bacillus, Klebsiella, Enterobacter*
Autotrophic	*Nostoc, Anabaena, Calothrix*
2. Symbiotic	*Rhizobium, Bradyrhizobium, Frankia, Nostoc, Anabaena*

can fix nitrogen are bacteria, some of which are free-living in soil and some in symbiosis with plants. The genera of organisms listed in Table 5.4 also include actinomycetes, e.g. *Frankia*, now classified alongside the bacteria.

The biochemistry of nitrogen fixation is similar in all organisms. Dinitrogen (nitrogen gas) is reduced to ammonia by the enzyme nitrogenase. (Enzymes are proteins, produced by living cells, which can catalyse specific biochemical reactions.) Nitrogenase contains iron and molybdenum and requires energy, which the microorganism obtains from the host plant, from soil organic matter, or from sunlight in the case of photosynthetic microorganisms. The source of energy depends on the particular microorganism as described below. A characteristic of the nitrogenase is that it is very sensitive to oxygen. The organism or its host therefore adopts strategies to exclude oxygen from the sites of nitrogenase activity.

Cyanobacteria

Because of their ability to photosynthesize and fix nitrogen the cyanobacteria (blue-green algae) are important in natural and agricultural ecosystems. They are early colonizers of rock debris and are believed to have been among the first colonizers of land two or three billion years ago. They are abundant in swamps and rice fields, and can fix nitrogen when free-living and in symbiosis with plants.

In wetland rice fields cells of *Anabaena* live in symbiosis with the tiny water fern *Azolla*. The cyanobacteria develop within the cells of the *Azolla*, which may protect it from excessive concentrations of oxygen. The symbiosis can fix about 50 kg N ha^{-1} a^{-1} and the material acts as fertilizer for the rice. Cyanobacteria also form a symbiosis with fungi, which is called a lichen, as referred to in Section 5.4.

Free-living heterotrophs

Several genera of bacteria are able to fix nitrogen. Under aerobic conditions *Azotobacter* and *Beijerinckia* appear to be the most common in soil. *Azotobacter*, which has been the most studied of the aerobic heterotrophs with this ability, requires a soil of pH 6–8, and as for all heterotrophs needs metabolizable organic carbon compounds. Under anaerobic conditions some species of *Clostridium* and other genera can also fix nitrogen (Table 5.4). Although the aerobic and anaerobic heterotrophs fix only small amounts of nitrogen, possibly 1 kg N ha^{-1} a^{-1} or less, they are probably important to the nitrogen economy of natural ecosystems. Inoculation of soil with *Azotobacter* has been tried but has been found to have little or no effect on the amount of nitrogen that is fixed.

Symbiotic nitrogen fixation

Symbiosis between nitrogen-fixing microorganisms and the host plant is important in providing nitrogen in natural plant communities and agricultural crops. Plant species with this symbiotic relationship often dominate as early colonizers of newly formed land.

Broadly, two types of symbiosis occur: rhizobia (*Rhizobium* and *Bradyrhizobium*) can form nodules (Figure 5.7) and fix nitrogen in the roots of members of the family Leguminosae; and actinomycetes classified in the genus *Frankia* can form nodules and fix nitrogen in 8 families of plants in which 17 genera are represented. The symbiosis between legumes and rhizobia has been intensively investigated in agricultural crops, for which it provides a valuable source of nitrogen.

Actinorhizal symbiosis occurs in tropical plants and trees of which *Casuarina* and *Ceanothus* are but two examples. In temperate regions it is perhaps best known in *Alnus glutinosa* (alder) and *Myrica gale* (bog myrtle). Of the several strains of *Frankia* that have been isolated, some are able to nodulate other plant species. *Frankia* occurs free-living in soil but has been mainly studied after isolation from root nodules. Its properties are still not well understood. The amounts of nitrogen fixed in actinorhizal symbiosis are difficult to measure, but increases of about 50 kg ha^{-1} a^{-1} in soils and standing timber have been found in stands of *Alnus* and *Casuarina*.

Less than 1% of the known species of legumes are grown commercially. Leguminous trees, bushes and plants undoubtedly contribute

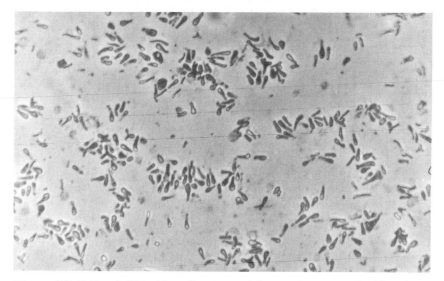

Figure 5.7 Cells of *Rhizobium* from a nodule on the roots of white clover (*Trifolium repens*). Within the nodule the cells become enlarged, as shown here, and are called bacteroids; each cell is 4 μm long. (Photograph P.J. Harris.)

nitrogen to their natural ecosystem through the nitrogen fixed by *Rhizobium* in their root nodules, although little is yet known about the amount of nitrogen they fix. Leguminous trees of the genus *Acacia* occur in much of the African savanna, and tropical forests are known to contain leguminous trees. One leguminous tree that has been investigated is *Leucaena leucocephala* because of its use in agroforestry in the tropics. Under ordinary field conditions it probably fixes about 100 kg N ha^{-1} a^{-1}.

The main legumes of agricultural importance are soya bean (*Glycine max*), *Phaseolus* species, groundnut (*Arachis hypogaea*), lucerne, also known as alfalfa (*Medicago sativa*), various clovers such as white (*Trifolium repens*) and subterranean (*Trifolium subterraneum*), and medics (*Medicago* species). The herbage legumes commonly fix between 100 and 200 kg N ha^{-1} and the grain legumes commonly fix between 40 and 100 kg N ha^{-1}. Their place in maintaining the fertility of soils is discussed in Chapter 13.

5.6 Ammonification and nitrification

Ammonification is the process of formation of ammonium ions from organic compounds (usually soil organic matter) that contain nitrogen.

Nitrification is the term applied to the oxidation of ammonium to nitrite and nitrate.

$$\text{Organic N} \xrightarrow[\substack{\text{many} \\ \text{organisms}}]{1} \text{NH}_4^+ \xrightarrow[\textit{Nitrosomonas}]{2} \text{NO}_2^- \xrightarrow[\textit{Nitrobacter}]{3} \text{NO}_3^-. \quad (5.1)$$

Ammonification (step 1) is brought about by a large number of heterotrophic microorganisms that derive carbon compounds and energy from soil organic matter. The NH_4^+ that is produced is the excess above their requirements for cell proliferation. Because of the wide range of microorganisms involved, the process can take place in soils that are too acid, too alkaline, too dry or too wet for rapid nitrification. Under such conditions NH_4^+ can be the main form of inorganic nitrogen in the soil. In most soils used for arable cropping NO_3^- is, however, the main product because ammonification is usually slower than nitrification (steps 2 and 3).

Nitrification is a stepwise process. In the above equation nitrite is given as the only intermediate, but it is probable that NH_2OH and NOH are also intermediates that exist only transitorily. Equations for the oxidation are:

$$\text{NH}_4^+ + 3\text{O} \rightarrow \text{NO}_2^- + \text{H}_2\text{O} + 2\text{H}^+; \quad (5.2)$$

$$\text{NO}_2^- + \text{O} \rightarrow \text{NO}_3^-. \quad (5.3)$$

It will be seen that nitrification is an acidifying process; the implications are considered more fully in Chapter 9.

Nitrosomonas and *Nitrobacter* are autotrophs and are believed to be the main genera of bacteria that bring about the oxidation of ammonium and nitrite, respectively. Some heterotrophs can cause nitrification but their effect is small except in acid soils.

Nitrification occurs most rapidly in moist soil (water potential of -10 kPa to -10^3 kPa) which is aerobic, has a pH of 5.5 to 8 and a temperature of 25–30 °C. It occurs more slowly in soil outside these optimum conditions. Above 10 °C in well-drained soils used for arable cropping, NH_4^+ is usually oxidized to NO_3^- within 3–4 weeks, whether the NH_4^+ comes from ammonification of organic matter or from fertilizers. In the tropics at about 25 °C and with other conditions optimal, NH_4^+ from application of fertilizer can be completely oxidized to NO_3^- within one to two weeks.

Nitrate can be found in soils of pH 4, possibly due to nitrification by heterotrophs. Autotrophic bacteria might also be active in such soils, either by adaption to acid conditions, or by occupying microsites at a higher pH than indicated by measurement of pH on several grams of soil. At pH above 8 the activity of *Nitrobacter* is reduced and NO_2^- can accumulate at concentrations toxic to higher plants.

Nitrification, being an oxidative process, does not occur under anaerobic (anoxic) conditions. Soils become anaerobic when they are waterlogged but only rarely is oxygen excluded from the whole soil profile. Nitrification will occur at microsites where oxygen is present and the nitrate may be reduced elsewhere in the soil. The process of nitrate reduction, known as denitrification, is discussed in the next section.

Nitrification can be inhibited by tannins, phenolic acids and phenolic glycosides that occur naturally in soils under some forest trees and may also occur in some grassland soils. There appears to be inhibition in soils under eucalyptus trees, but evidence for inhibition by specific naturally occurring chemicals is inconclusive. Added chemicals, including several soil fumigants, and herbicides and pesticides applied at excessive rates, stop nitrification. Various chemicals have also been investigated as inhibitors in order to prevent the formation of nitrate, which can be lost by leaching and denitrification; two common ones are nitropyrin (2-chloro-6-(trichloromethyl)pyridine) and dicyandiamide (see Table 8.9).

5.7 Denitrification

This term refers to the reduction of NO_3^- or NO_2^- to N_2 and oxides of nitrogen by microbial activity (biological denitrification), and to the chemical reduction of NO_2^- and other unstable nitrogen compounds (chemical denitrification). Biological denitrification is often called dissimilatory denitrification to distinguish it from assimilatory denitrification in which microorganisms assimilate and reduce nitrate as a first step in protein synthesis. The process of denitrification is a stepwise reduction:

$$NO_3^- \to NO_2^- \to X \to N_2O \to N_2, \qquad (5.4)$$

where the intermediate X might be nitric oxide, NO, though this is not certain. Many bacterial genera possess enzymes for each step in the reduction of nitrate to nitrogen gas. Some take the reduction only to nitrite, and others may be unable to reduce nitrous oxide. Many genera with the capability to denitrify are found in soils, but it is difficult to know whch are the most active. Species of *Pseudomonas* and *Alcaligenes* have commonly been isolated and bacteria of both genera are known to denitrify.

Most denitrifying bacteria are facultative anaerobes; they use organic substances as a source of energy and are therefore heterotrophs. In aerobic conditions they use O_2 as the electron acceptor (Section 5.8) but can adapt to the absence of oxygen, or to low concentrations of oxygen, by using oxygenated forms of nitrogen, for example nitrate, as alternative electron acceptors, which are thereby reduced (see Equation 5.10).

The reduction of nitrate to nitrogen gas requires a series of enzymes. The enzyme that reduces nitrate to nitrite is called nitrate reductase, and nitrite is reduced in turn by nitrite reductase. In the course of reduction the gases nitric oxide, NO, and nitrous oxide, N_2O, are evolved and diffuse into the atmosphere. Under the most reducing conditions the main product is nitrogen gas. Nitrous oxide is implicated in global warming (Chapter 11), and soil is its main source.

Nitrous oxide reductase has been found to be inhibited by acetylene. Use has been made of this observation to measure denitrification. By blocking the final step in the reduction, total denitrification can be measured as N_2O production, which is possible with sensitive gas chromatographs, whereas measurement of comparatively small amounts of N_2 is almost impossible because of its high background concentration in the atmosphere.

Denitrification requires the presence of nitrate, metabolizable carbon compounds and the almost complete absence of oxygen at the site of reduction. Nitrate is formed by nitrification or might be added in fertilizer. Soil organic matter, plant roots and organic manures provide metabolizable carbon compounds. The concentration of oxygen is reduced to a sufficiently low level when the soil air is displaced by water, as after heavy rainfall or irrigation, or flooding. The soil does not need to be devoid of oxygen because denitrification will occur at microsites that are anaerobic, for example within water-saturated aggregates or where an energy-rich substrate causes oxygen depletion, even though the soil as a whole contains oxygen.

The rate of denitrification increases with temperature and is highest at a soil pH between about 6 and 8. Under more acid conditions not only is it slower but the ratio of $N_2O : N_2$ is higher, either because of chemical reduction of nitrite to N_2O or inhibition of N_2O reductase. It has also been found that nitrate inhibits N_2O reductase, which results in a higher ratio of $N_2O : N_2$ in its presence.

5.8 Oxidation and reduction

Nitrification and denitrification, discussed in Sections 5.6 and 5.7, are examples of oxidative and reductive reactions, respectively. Soil microorganisms carry out many more reactions of these two kinds. An important reaction is the oxidation of plant residues, which can be illustrated with glucose as the starting substance.

The oxidation occurs stepwise using a series of enzymes, with pyruvic acid as the product of the first step. The reaction is

$$C_6H_{12}O_6 \rightarrow 2CH_3COCOOH + 4H^+ + 4e^-, \qquad (5.5)$$

where e^- represents one electron. There is a further reaction:

$$O_2 + 4H^+ + 4e^- \rightarrow 2H_2O, \qquad (5.6)$$

in which O_2 gains electrons and is reduced to water.
The overall reaction may be written

$$C_6H_{12}O_6 + O_2 \rightarrow 2CH_3COCOOH + 2H_2O. \qquad (5.7)$$

In the absence of O_2, other substances gain electrons and are reduced, as is oxygen in Equation 5.6. Examples are:

$$Fe(OH)_3 + 3H^+ + e^- \rightarrow Fe^{2+} + 3H_2O; \qquad (5.8)$$
$$MnO_2 + 4H^+ + 2e^- \rightarrow Mn^{2+} + 2H_2O; \qquad (5.9)$$

and for the reduction of nitrate

$$NO_3^- + 2H^+ + 2e^- \rightarrow NO_2^- + H_2O; \qquad (5.10)$$

and nitrite

$$2NO_2^- + 8H^+ + 6e^- \rightarrow N_2 + 4H_2O. \qquad (5.11)$$

In these reactions there is reduction of oxygen or of another substance when the reaction is from left to right, and oxidation (where it occurs) is in the reverse direction. Oxidation produces protons and reductive reactions consume them. This is discussed further in Chapter 9, which deals with soil acidification.

In aerobic conditions oxidation proceeds beyond that shown by Equation 5.7, ultimately with the formation of carbon dioxide and water:

$$C_6H_{12}O_6 + 6O_2 \rightarrow 6CO_2 + 6H_2O. \qquad (5.12)$$

With substances other than oxygen as electron acceptors, oxidation is usually incomplete and various organic compounds are produced including methane, ethylene and acetic acid. All these oxidation/reduction reactions are driven by the soil microorganisms.

The state of oxidation/reduction can be measured as the difference in voltage between a platinum and a calomel electrode placed in the soil. The voltage difference is known as the *redox potential*, (E_h). In a neutral, well-aerated soil it is over 500 mV, but when soil becomes anaerobic the redox potential falls. Nitrate is readily reduced; that is, it gains electrons. It does this more readily than Fe^{3+}, for example, and is reduced at a higher redox potential. As the soil becomes more anaerobic the demand for alternative electron acceptors increases, and reduction occurs in the order: $O_2 > NO_3^- > Mn^{4+} > Fe^{3+} > SO_4^{2-} > CO_2 > H^+$.

However, soil is never completely homogeneous and oxygen may still be present in some pores when reduction of other substances is occurring elsewhere.

Because oxidation/reduction processes are driven by the activity of microorganisms, they are most pronounced at high temperatures. Hence, methane is an important gaseous product under anaerobic conditions in hot countries, as in wetland rice soils; as a greenhouse gas it is referred to in Section 11.7.

5.9 Summary

The soil provides diverse habitats for organisms (the soil fauna and microorganisms). Their numbers (bacteria in the top few centimetres of soil usually number between 1 million and 100 million per gram of soil) and activity depend to a large extent on the addition of organic matter to the soil as leaf litter and plant residues, and from the decay of roots.

The microorganisms include bacteria, actinomycetes, fungi, algae and protozoa. Most require an organic substrate whereas others do not; some require a supply of oxygen, although others do not. Some species oxidize inorganic substances, for example ammonium to nitrite, or nitrite to nitrate. Others can convert atmospheric nitrogen to organic nitrogen compounds. Some bacteria reduce oxidized substances, producing, for example, nitrous oxide, nitric oxide and methane, which pass into the atmosphere.

Because of their diverse activities soil organisms have an important role in the environment and in crop production.

6

Movement of water, air, solutes and heat in soil

6.1 Introduction

The soil components described in Chapter 2 include solids (minerals and organic matter), liquids (water) and gases (the soil air). Of these, water can readily be seen to move when rain falls on dry soil. The water wets the soil surface and if a vertical face is cut into the soil the wetting front can often be observed as a colour change. If more rain falls and the soil is free-draining the water passes into drains or percolates through the subsoil and may move into underlying porous rock. If the rock is not porous the water will emerge as surface seepage on hill slopes and run into rivers.

The soil solids can also move, most obviously when they are blown by wind or washed down slopes by rainwater. They also move *in situ* during freezing and thawing, during swelling and shrinking, and by the activities of the soil fauna, especially earthworms and termites. The flow of gas into and out of soil cannot be observed by eye (the gas bubbles that emerge from swamps are an exception), but its occurrence can be demonstrated by collecting carbon dioxide that is evolved from soil in a closed flask.

Soil also has thermal energy, which we refer to as heat, changes in which manifest themselves as changes in temperature. The soil surface gains heat from incoming solar radiation and loses it by back-radiation and convection to the atmosphere, by evaporating water and by transfer down the profile. Solutes also move: they diffuse and they are carried in the flow of water.

Soil is never truly in equilibrium with its environment although we often assume an equilibrium state in order to develop an understanding of processes. Most of the time there is movement of water, gases, solutes and heat into, out of, and within the soil.

In this chapter we consider the movement of four components: water, gases, solutes and heat. They are considered together because they have characteristics in common, although important differences will be pointed out. The movement of solids is quite different and is discussed in Chapter 12.

6.2 Principles

The two mechanisms by which components of the soil move are mass flow and diffusion.

The more familiar to us is mass flow, which is the movement of fluids (gases and liquids) caused by a force or pressure difference, e.g. water flowing downhill or air being pumped into a tyre. During such movement work is done in overcoming frictional resistance and usually in changing the energy of the fluid. When water flows downhill it loses potential energy, owing to loss of height, and temporarily gains kinetic energy; air pumped into a tyre gains energy from the input of work. A characteristic of mass flow is the movement of the whole phase, that is, dissolved components move at the same rate as the water, and in mass flow of air all the component gases move at the same rate.

The other mechanism of movement, diffusion, is due to the random thermal motion of molecules and ions, no energy change being involved. Diffusion occurs whenever there is a difference in concentration of a substance between two points. If there are more molecules or ions of a substance at one point than at another, more will be moving away from that point by their thermal motion than will be moving towards it. Hence there is a net movement of the substance down a concentration gradient until its concentration becomes uniform. The rate depends on the velocities of the particles (e.g. molecules), on the concentration gradient and on the cross-sectional area available for diffusion. One difference from mass flow is that in a mixture of components that diffuse, for example the gases in air, movement of the components may differ according to their concentration gradient and thermal motion of their molecules. As will be seen later, the conduction of heat is exactly analogous to diffusion.

We start with the flow of water. In many domestic water systems a tank in the roof space acts as a reservoir that supplies water to the taps. Water flows by gravity down pipes to the taps because the water in the reservoir has a larger gravitational potential than at the outlet from the taps. (Potential is the ability to do work and when concerned with water movement is expressed as the height of an equivalent water column; see Section 2.4.) When dealing with soil water, the symbol used for potential is ψ (Greek psi), and the difference in potential in the example given is expressed as $\psi_{tank} - \psi_{tap}$. The rate of flow of water increases as ($\psi_{tank} - \psi_{tap}$) increases, but becomes less as the length of the flow path (l) is increased, that is

$$q_w = K(\psi_1 - \psi_2)/l, \qquad (6.1)$$

where q_w is the flow rate per unit area. The flow rate also depends on the properties of the conducting pipe, particularly on its diameter, and this will be reflected in the value for the constant K in the equation. ($\psi_1 - \psi_2$)/l is the gradient of the potential, which can be regarded as the 'driving force' for flow. For small values of ($\psi_1 - \psi_2$) and l, the gradient is expressed as $d\psi/dl$. Since flow is in the opposite direction to the gradient, that is, water flows in the direction of decreasing potential, a negative sign is required in Equation 6.1, which becomes

$$q_w = -K \, d\psi/dl. \qquad (6.2)$$

This is one form of an equation known as Darcy's Law.

Application of Equation 6.2 to soil will be considered by reference to Figure 6.1, which shows a cylinder of water-saturated soil open at the bottom and with a thin layer of water on the top. Ignoring the effect of the water layer, the gradient of the gravity potential is ($\psi_1 - \psi_2$)/($x_1 - x_2$), or $d\psi/dl$. For this condition, where gravity is the only force acting on the water, and expressing each potential (ψ_1 and ψ_2) as the height of an

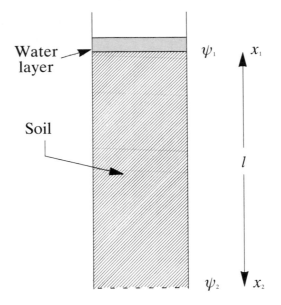

Figure 6.1 At the top of the cylinder of soil saturated with water there is a greater gravity potential, ψ_1 than at the bottom, ψ_2. The difference in potential is $\psi_1-\psi_2$ and the gradient is $(\psi_1-\psi_2)/(x_1-x_2)$.

equivalent water column, $(\psi_1-\psi_2) = (x_1-x_2)$, and hence the gradient of $d\psi/dl = -1$). In this example, therefore, $q_w = K$, where K is the hydraulic conductivity of the soil and reflects its ability to conduct water. In saturated soil the value of K is known as the saturated hydraulic conductivity, or K_{sat}. This property and the conductivity of unsaturated soils for water are discussed in the next section.

The form of Equation 6.2 is the same as that for the diffusion of gases. Diffusion occurs in a volume of mixed gases if their concentrations differ between any two points. Suppose a volatile liquid is placed at the bottom of an open-ended cylinder (Figure 6.2). Immediately above the liquid its saturated vapour pressure is known and its concentration, C_s, can be calculated. If the concentration at the open-end is C_o, the concentration gradient when a steady state is reached is $(C_s-C_o)/l$ where l is the length of the column. If C_o is small where the concentration of the vapour mixes with the atmosphere the concentration gradient is approximately C_s/l. The rate of loss of vapour from the cylinder is then described by

$$q_g = -DC_s/l, \tag{6.3}$$

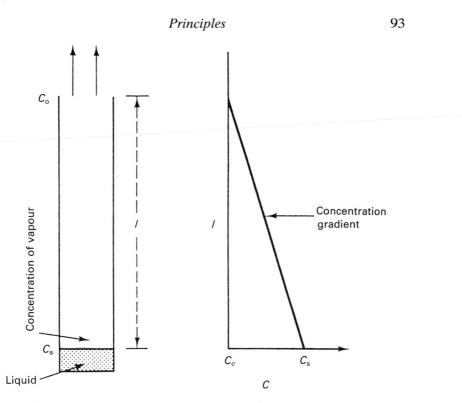

Figure 6.2 Representation of steady-state diffusion of vapour.

and the general form is

$$q_g = -DdC/dl, \tag{6.4}$$

where q_g is the rate of loss of mass of the vapour and D is the diffusion coefficient of the vapour (see also Equation 6.8).

Equation 6.4 is a form of Fick's first law, and can be most simply applied to one-dimensional diffusion where there is a linear concentration gradient, as in the example above.

Diffusion in gases occurs because the molecules, being in constant motion, tend to spread uniformly throughout the available space. Molecules and ions in solution behave similarly. This rate of movement in solution is described by an equation of the same form as Equation 6.4:

$$q_s = -DdC/dx, \tag{6.5}$$

where q_s is the flux of solute across unit area in unit time, x is distance, and dC/dx is the concentration gradient; D is the diffusion coefficient of the molecules or ions in solution.

The conduction of heat is described by equations that are mathematically analogous to those given above:

$$q_h = -k \, dT/dx, \qquad\qquad (6.6)$$

where q_h is the amount of the heat conducted across unit area in unit time, dT/dx is the temperature gradient and k is the thermal conductivity.

Equations 6.2–6.6 can be simply applied to movement in one dimension under steady conditions, that is, where the gradient of concentration or potential is linear and remains constant in time. Often, the flux occurs in more than one dimension and the concentration (or potential) gradient is neither linear nor constant, as for the diffusion of solutes away from a fertilizer granule, and for movement of water out of furrows carrying irrigation. Mathematical solutions have been obtained for many of these conditions but are beyond the scope of this book.

We can now consider the flow of the individual components in more detail.

6.3 Movement of water

As discussed above (see also Section 2.4) the downward flow of water in a column of saturated soil is due mainly to the gradient of gravitational potential. In unsaturated soils, surface tension and adhesion of water to soil surfaces give rise to another potential known as the matric potential. Also, dissolved ions and molecules give an osmotic potential (Section 2.4). In general, water moves from high to low potential, whether the potential is due to gravity, capillarity or osmosis, or to their combined effect.

In unsaturated soils water moves down a gradient of matric potential (ignoring any contribution from gravity or osmosis). The water creeps over the solid particles from where the films are thickest to where they are thinner. Consider, for example, a portion of soil at field capacity (suction 0.5 m = potential −5 kPa) in contact with a portion of the same soil at the permanent wilting point (suction 150 m = potential −1.5 × 10^3 kPa). Water will move down the gradient of potential to the portion of soil at the permanent wilting point which has the lower (more negative) potential. In this example water moves from wet to dry soil, which might be expected, but there are exceptions. Water does not normally move from clay soil to sand, which may be much drier, because the forces holding the water in the clay are stronger than those holding it in the sand. At the same water content the matric potential (due to surface tension) in the clay is less (more negative) than that in the sand. The direction of water movement, in other words, is determined by difference in total water potential rather than by difference in water content.

The rate of flow of water described by Equation 6.2 depends partly on K, the hydraulic conductivity, which has the same units as velocity, for example, cm d^{-1} or m s^{-1}. K is a soil property and is largely determined by the size and continuity of the pores. Poiseuille's Law states that the flow rate in a capillary tube is proportional to the fourth power of the radius (r^4). In saturated soil, where all pores are filled with water, the flow rate is high in coarse sands, soil composed of aggregates, and where channels and cracks are present. The value of hydraulic conductivity in saturated soils is known as K_{sat}. It can be as low as 1 mm per day in compacted clay and 10 000 times higher (10 m per day) in a coarse sandy soil.

As soils become dry, water is strongly held at low potentials and is restricted to narrow pores and thin films. Under these conditions the rate of water movement decreases very greatly because of the large resistance to flow in narrow pores. This is illustrated by the relation between water content and the hydraulic conductivity shown in Figure 6.3, where K changes by five orders of magnitude over the range of water content likely to be observed in the field. Hence the rate of water flow in unsaturated soils tends to be very slow relative to that at saturation.

6.4 Infiltration and percolation of water

Infiltration refers to the entry of water into the soil; percolation is the movement of water through the soil profile.

When rain falls on newly cultivated, dry soil its infiltration rate is high because of the presence of wide pores and the low water potential of dry soil. The rate decreases as the structure collapses and the water potential rises, and then the rate gradually reaches a lower, steady value. This steady value is called the percolation rate and is the same as K_{sat}, referred to above, if all the pore space is filled with water and the soil is deep and uniform with depth.

Infiltration recharges the soil profile after a dry period. After a single rainfall event has wetted the top one or two centimetres of soil, downward movement of water is mainly along the gradients of gravitational

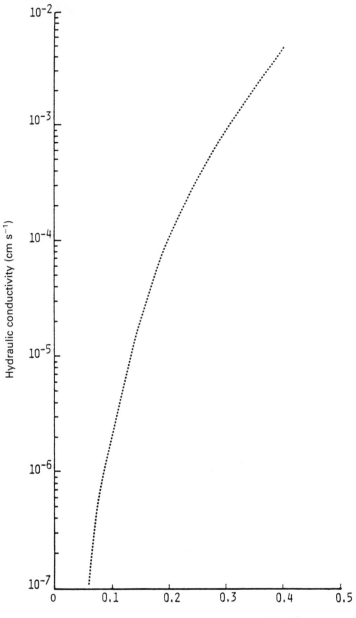

Figure 6.3 The relation between water content of soil and hydraulic conductivity. Note the logarithmic scale and the very slow rate of flow as the soil becomes less wet. (From Hillel, D. 1980. *Fundamentals of Soil Physics*. Academic Press, New York; with permission.)

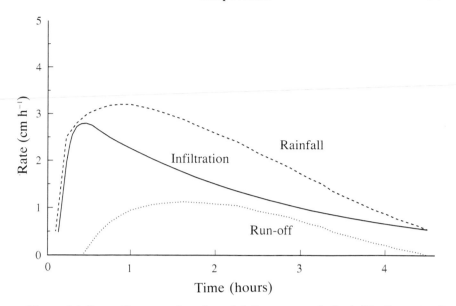

Figure 6.4 Run-off occurs when the rainfall rate exceeds the infiltration rate of water into the soil.

and matric potential. The rate of flow falls sharply as the water films become thin, often with a clearly visible water front. This front disperses only slowly as a result of very slow movement of liquid water into the drier soil beneath or by diffusion of water vapour.

The rate of infiltration is important in relation to the intensity of rainfall. When rainfall intensity is greater than the rate of infiltration the excess rain runs off the soil surface (Figure 6.4); this is one of the causes of water erosion (Chapter 12). In order to minimize run-off the soil surface should contain wide pores or cracks and the structure should be stable. Wide pores and cracks also conduct water to drains where these are installed, or to the subsoil, or laterally through soils on a hill slope. Maintenance of these conducting channels is a key requirement in the management of many soils.

6.5 Evaporation of water from soil

Liquid water has a vapour pressure, which increases with temperature. At constant temperature, vapour passes from the liquid into an enclosed volume of dry air at a rate that decreases with time until the air becomes saturated. When the air is saturated, transfer rates of water from and to the liquid are equal; the relative humidity is 100% and the deficit from saturation is zero.

Evaporation will, however, continue if unsaturated air blows over the surface. It will also continue if there is a temperature gradient, vapour diffusing from a warm to a cold surface where the vapour pressure, and hence concentration, is lower. When air cools at night the relative humidity increases, and if it reaches saturation in the cooled air some water vapour condenses as fog and dew.

There are three requirements for evaporation:

(i)   a supply of energy to meet the latent heat of vaporization; over a short period this may come from cooling of the soil, but prolonged evaporation requires an external source of energy of which solar radiation is usually dominant;

(ii) a gradient of vapour pressure such that the vapour pressure at the site of evaporation is greater than that of the overlying air;

(iii) a means of conveying vapour from the relatively wet air at the site of evaporation to the relatively dry, overlying air. In perfectly still air an uncovered surface will lose vapour by diffusion, but turbulent transfer of vapour in moving air, which depends on wind speed, is usually the dominant means of vapour transport.

Provided the air is unsaturated and water is freely available, condition (ii) will be satisfied because the air in contact with the evaporating surface will be saturated with water vapour. Because the air at the surface of a wet soil is saturated, evaporation will be similar to that from a free water surface, and is called potential evaporation. If, however, the upward movement of water through the soil is less than that required to maintain potential evaporation, the soil surface will dry, causing the vapour pressure in the soil air to decrease. This decrease may be sufficient to reduce the vapour pressure gradient between the soil air and the overlying air to a very low value so that evaporation becomes very slow.

It is evident from the conditions stated above that the potential evaporation from a wet surface depends on the energy input from solar radiation, the vapour pressure of the air, the saturated vapour pressure of the surface (which depends on temperature) and the wind speed. All of these can be known from routinely collected meteorological records. To predict evaporation from a drying soil, however, information is needed about the factors controlling the movement of water through the soil.

6.6 Transpiration by plants

The principles are the same as set out in Section 6.5. There are, however, two important differences between evaporation from within leaves (transpiration) and evaporation from soil. First, in transpiration vapour in the intercellular spaces in leaves has to diffuse to, and through, the stomata of the leaf surface before reaching the air. This extra barrier to water movement restricts evaporation from a well-hydrated leaf to, typically, 80% of that from wet soil. Secondly, evaporation from soil is reduced by shortage of water only when the soil surface is sufficiently dry for the vapour pressure in the soil air to be reduced substantially below the value at saturation. If a leaf were to reach such a state of desiccation there would be irreversible physiological damage, so plants have developed a means of restricting evaporation that prevents severe desiccation.

On a very hot dry day a complete canopy of plants growing in wet soil transpires up to about 10 mm of water per day, although 3–5 mm is more usual for hot summer days in a temperate climate. Evaporation decreases the leaf water content, thus lowering the water potential of the leaves. Because water moves from high to low potentials, water then moves from roots via the xylem to the leaves. This in turn lowers the water potential in the roots, resulting in an inflow of water from the soil along a gradient of water potential. If the soil is unable to supply water at a sufficient rate, leaves dehydrate further, thereby lowering their water potential. When dehydration is severe, the stomata close in order to reduce transpiration, the leaves become warmer because less heat is being lost through the evaporation of water, and they may wilt. The leaves may recover when the evaporative demand and hence transpiration decreases at night, but many plant species become desiccated and die if the leaves are wilted for more than a few days.

Estimating transpiration loss is important for water engineers who need to know the fraction of the rainfall that will be lost by transpiration and hence the fraction that will eventually flow into rivers or replenish aquifers. It is also important in scheduling irrigation (Section 8.10). Calculated losses have been found to agree with those from short grass and crop canopies. The loss is greater from trees because humid air in the immediate vicinity of leaves is more rapidly mixed with the drier air in the atmosphere above. For isolated plants or trees, or small plots of crop plants, the loss is also greater because they are surrounded by dry air.

The effect of transpiration is to reduce soil water content to a greater depth than that due to evaporation alone. The greater the depth of the root system the greater the amount of water that is available to the plants and the greater the depth to which the soil is dried. The uptake of water by plants is discussed in Section 7.4.

6.7 Diffusion of gases

Soil air differs in composition from that of the atmosphere in containing higher concentrations of carbon dioxide and water vapour and less oxygen (Section 2.6). Carbon dioxide and water vapour therefore move out of soil into the atmosphere and oxygen moves into the soil. Movement is by diffusion because of the concentration gradient, alternatively described as a gradient of partial pressure. Equation 6.4 describes the rate of transfer.

The diffusion coefficient of gases in soil is smaller than in air, partly because the soil air occupies only a fraction, e_a, of the cross-sectional area. Another reason is that the soil pores are not vertical, parallel tubes. This means that the diffusing gas molecules have to take a longer and more tortuous path than the straight line distance between two points. The path length, and hence the tortuosity, is not constant but increases as more pores become water-filled. For a dry soil, an equation developed by Penman that may be used to describe the relationship between the diffusion coefficients in soil, D_s, and in air, D_o, is

$$D_s = 0.66\, e_a\, D_o, \tag{6.8}$$

where e_a is the air-filled porosity.

Within the soil, diffusion of oxygen to and carbon dioxide away from microorganisms and plant roots occurs through water films. Both gases are soluble in water, carbon dioxide being about 25 times as soluble as oxygen. Diffusion is very much slower in liquids. For oxygen the value of D is about 2×10^{-5} m^2 s^{-1} in air compared with 2×10^{-9} m^2 s^{-1} in water (Table 6.1). This, together with the low solubility of oxygen, means that a 1 mm thick water film is as effective a barrier as 300 000 mm of air. Hence a very thin water barrier can limit the supply of oxygen to microorganisms especially those inside soil aggregates. Depending on the rate of respiration, aggregates with a diameter bigger than 1–2 cm will have an anaerobic core when their internal pore space is full of water.

The total pore space and water content of soil have dominant

Table 6.1. *Some typical diffusion coefficients*

	$T(°C)$	$D(m^2 \, s^{-1})$
Gas phase		
O_2 in air	25	2.1×10^{-5}
CO_2 in air	25	1.6×10^{-5}
Liquid phase		
O_2 in water	25	2.26×10^{-9}
CO_2 in water	25	1.66×10^{-9}
NaCl in water	25	1.61×10^{-9}
Soil		
Cl in sandy clay loam at 20% water by vol.	lab	2.4×10^{-10}
PO_4 in sandy clay loam at 20% water by vol.	lab	3×10^{-14}

Source: From Nye, P.H. and Tinker, P.B. 1977. *Solute Movement in the Soil-Root System.* Blackwell Scientific Publications, Oxford, with permission.

influences on the rate of gas diffusion. The water content affects tortuosity, as mentioned above, but usually the main effect is due to the low air-filled porosity, e_a, when the soil has a high water content (see Section 2.3). If the total pore space is completely occupied by water, oxygen transfer usually depends on the slow diffusion through water, which is insufficient for most plants. There are exceptions: in some plants the transfer of oxygen is internally to their roots through air-filled pore spaces (Section 7.3). In most conditions, oxygenation of water-logged soil is restricted to the surface few millimetres.

In addition to diffusion as described above, mass flow of gas can occur between the soil and the atmosphere. It occurs when there are changes of temperature and pressure and it can be induced by air turbulence over the soil surface. In addition, the ingress of rain and irrigation water displaces some of the soil air and draws in air from the atmosphere as the water moves downwards. Although mass exchange occurs, diffusion is the far more important process of movement.

6.8 Movement of solutes

Ions and molecules dissolved in water constitute the soil solution (Section 2.5). The concentration and composition of the solution changes when (i) mineralization of soil organic matter releases nutrients, (ii) fertilizers and pesticides are applied, (iii) ions are taken up by plants, and (iv) rainfall and irrigation increase, and evaporation/transpiration decreases, the soil water content. The result of these processes is spatial variation in the concentration and composition of the solution, which gives rise to diffusion. The movement of solutes is also caused by the flow of water carrying solutes with it, a process known as mass flow or convective flux. Two processes of solute movement therefore need to be described; we start with diffusion.

The rate of movement of solutes by diffusion in free solution is described by Equation 6.5. Diffusion across unit cross-section of soil is slower than across unit cross-section of liquid because only part, θ_v, of the cross-sectional area is occupied by liquid (as with diffusion of gases through air-filled pore spaces discussed above). The diffusion coefficient of a solute in soil is therefore less than in free solution and can be written as D_s, to distinguish it from the coefficient in free solution, D_o. As with gases, diffusion occurs through pores which are not all oriented in the direction of the concentration gradient. The diffusive path is therefore longer, which reduces the diffusive flux by the tortuosity factor, f, which becomes smaller the drier the soil.

The diffusive flux of solutes in soil is also influenced by the adsorption/desorption properties of soil. Adsorption reduces the rate at which solutes diffuse through soil: the greater the adsorption (or buffer capacity, b, of the soil for the solute), the slower the diffusion.

The diffusion coefficient of a solute in soil, D_s, is therefore less than in free solution, D_o, because of the lower cross-sectional area for diffusion, the tortuosity factor for the diffusive pathway and the buffer capacity. Hence

$$D_s = \frac{D_o\,\theta_v\,f}{b}. \tag{6.9}$$

The values of D_o for small ions in water are all of the same order of magnitude. The values of D_s differ considerably between ions because of the effect of the buffer capacity, b (Table 6.1). For example, chloride is not adsorbed and diffuses much more rapidly in soil than phosphate, which is strongly adsorbed.

Mass flow or convective flux of solutes occurs in water flow, discussed in Section 6.3. The discussion followed the usual practice of describing this as water flow although in reality it is flow of the soil solution.

If the solution flows through a portion of soil with which it is in equilibrium with respect to the solutes it contains, its composition will not change. If the flow rate is V and its concentration of ions is C_1, the flux, F, of ions is given by

$$F = V C_1. \tag{6.10}$$

Often the soil is not in equilibrium with respect to dissolved salts. This occurs, for example, when fertilizer is added or soil organic matter mineralizes. The effect can be demonstrated by adding a salt solution to the top of a column of soil and then slowly adding water. The soil behaves like a chromatographic column in which adsorbed ions move more slowly than the water. According to chromatographic theory the retardation factor, RF, is given by

$$RF = \frac{1}{1 + (b\rho/\theta)}, \qquad (6.11)$$

where b is the adsorption coefficient in $cm^3\ g^{-1}$. To convert the adsorption parameters to those in the soil column the equation includes the dry bulk density of the soil, ρ, and the water content of the column, θ.

When mass flow occurs, it usually transports solute much faster than does diffusion. Pesticide molecules and most ions are, however, adsorbed, and this can slow down transport by mass flow considerably. Borate is weakly adsorbed; if water moves 1 m through a soil profile, the borate ions move only about half as far. Under the same conditions phosphate ions might move only 2 mm.

6.9 Soil temperature and movement of heat

Soil temperature has important effects on the activity of soil organisms, germination of seeds, seedling emergence and root growth. It strongly affects the processes of rock weathering and soil formation. It is also one of the factors that determines evaporation of water, water vapour tending to move down a temperature gradient. The rates at which chemical reactions and biological processes occur generally increases two- or three-fold for each 10 deg rise in temperature (within a limited range).

Soil temperature may differ from air temperature by several degrees. The difference is due to four factors:

incoming solar radiation and the soil albedo;
heat loss by radiation, convection, conduction, and latent heat of vaporization of water;
thermal conductivity of soil;
heat capacity of soil.

At a particular site the incoming radiation rate depends on latitude, angle of the surface to the radiation, the season of the year, and cloudiness. The effect of season is shown by measurements at Rothamsted

Table 6.2. *Albedo of some terrestrial*
surfaces expressed as a percentage of
incoming radiation

Surface	Albedo (%)
Fresh snow	75–95
Dry sand	35–45
Dark, bare soil	5–15
Light, bare soil	20–35
Meadow grass	10–20
Savanna	15–30
Deciduous forest	10–20
Coniferous forest	5–15

Source: From Taylor, S.A. and Ashcroft, G.L. 1972. *Physical Edaphology.* Freeman, New York; with permission.

Experimental Station in southern England where the received daily radiation R, averaged over 13 years, was 1.5 MJ m^{-2} in December and 17.6 MJ m^{-2} in June.

Part of the solar radiation is reflected by soil, rocks, water and leaves of plants. The percentage or fraction that is reflected is known as the albedo, α, of the surface; values are given in Table 6.2. The remaining part of the radiation, $R(1-\alpha)$, is absorbed at the soil surface where it is changed into heat, which warms the soil and the air above the soil and vaporizes water. An equation for the energy balance can be written:

$$R(1-\alpha) = S + A + LE, \tag{6.12}$$

where S is the heat flux into the soil, A is heat flux into the air above the soil (including convection and back-radiation), E is the rate of evaporation of water and L is the latent heat of vaporization. A soil which is wet at the surface loses water at about the same rate as a free water surface; under these conditions loss of heat by evaporation of water is high. When the surface dries during a sunny day, loss of water (and hence of latent heat) decreases so that the soil and the air above it become warmer.

The extent of the temperature rise in a soil depends on its thermal conductivity and heat capacity. The heat flux in the soil is described by Equation 6.6:

$$q_h = -k \, dT/dx, \tag{6.13}$$

Table 6.3. *Thermal properties of the soil constituents*

	Volumetric heat capacity, C (kJ m^{-3} deg^{-1})	Thermal conductivity, k (W m^{-1} deg^{-1})
Air	1.2	0.025
Water	4.2×10^3	0.6
Quartz	2×10^3	8.8
Clay minerals	2×10^3	2.9
Soil organic matter	2.7×10^3	0.25

where k is the thermal conductivity of the soil. The values of k for the main soil components (Table 6.3) show that quartz, the dominant mineral of most sands, is the best conductor of heat and air is the poorest, that is, air is a particularly good insulator.

The volumetric heat capacity, C, the amount of heat required to raise unit volume by one degree, also differs greatly between the soil components (Table 6.3). Air requires the least amount of heat to raise its temperature by 1 deg and water requires most. Hence for a given input of heat the increase of temperature is greater when soils are dry than when they are wet. Additionally, wet soils lose heat as latent heat of vaporization, also they are better conductors, leading to more heat loss to the subsoil.

The rate of change of temperature with time, dT/dt, at any depth in the soil will be seen to depend directly on the thermal conductivity, k, and inversely with the volumetric heat capacity, C, that is, dT/dt varies as k/C, a quotient known as the thermal diffusivity. On a sunny day when the soil surface has become warm, there is heat flux into the soil and at night the soil surface loses heat. The fluxes are greater for wet soils than for dry; indeed, irrigating soil has been suggested as a means of night-time protection of frost-sensitive crops such as grapes because of the increased heat output at night.

A freely draining moist sand warms quickly to depth during the day and cools quickly at night, in contrast to a wet soil which shows less temperature variation between night and day. In addition to the thermal conductivity and volumetric heat capacity of the soil components, the transmission of heat depends on the area of contact between the solid and liquid components, air being a poor conductor. As a consequence, dry sand conducts heat more slowly than sand containing a little water because the water increases the area of contact between the sand grains.

The temperature regime of soil is strongly influenced by vegetation.

Diurnal and seasonal fluctuations of temperature are reduced because the vegetation intercepts part of the solar radiation and the back-radiation from the soil. The air within a plant canopy is a poor conductor of heat, and this also damps the temperature fluctuations of the soil beneath. A layer of snow and a mulch of dead vegetation have a similar effect. Under a mulch, for example, the soil temperature is lower during a sunny day and higher at night than that of a bare soil.

6.10 Summary

Water, gases, solutes and heat move into, out of, and within the soil, which is never in true equilibrium with its environment. The processes of movement are (i) mass flow caused by a force or pressure difference, and (ii) diffusion, which is caused by a difference of concentration for gases and solutes, and temperature for heat.

Water moves into the soil from rainfall by infiltration and is returned to the atmosphere by evaporation and transpiration from plants. Movement within the soil becomes very slow as the soil dries.

Soil air contains less oxygen and more carbon dioxide than the atmosphere because of respiration by soil microorganisms and roots; the difference in concentrations results in diffusion.

Solutes in the soil solution move by mass flow when the liquid phase moves, as to plant roots or ground water. They also move by diffusion when the soil solution close to plant roots becomes depleted by plant uptake, and when fertilizer is added.

Soils receive radiation from the sun, which heats the surface of the soil. The thermal conductivity and volumetric heat capacity of the soil components, and the loss of heat as latent heat of vaporization of water, determine the effect of the received radiation on soil temperature.

B
Soils in relation to the environment

7

Soil as a medium for plant growth

7.1 Introduction

Although plants can be grown to maturity in aerated nutrient solutions and other media, virtually all land plants, including those on which we depend for food, fibre and fuel, grow in soil. The following requirements of plants can be met by soil:

anchorage for roots;
supply of water;
supply of air and particularly oxygen;
supply of mineral nutrients;
buffering against adverse changes of temperature and pH.

Soils rarely provide ideal conditions for plant growth. For the growth of crops, soil properties are usually changed in order to ensure good yields, most crop plants having been selected and bred for high production in fertile soil. In contrast, wild plants are adapted to the local conditions, including the climate, and possibly to low supplies of water, oxygen and nutrients, extreme pH, waterlogging, high concentrations of toxic elements, competition from other species or little anchorage.

In this chapter the emphasis is on principles, illustrated mainly by reference to crop plants, which have been more intensively studied than wild plants. For readers not familiar with the subject, the first section is an outline of the development and growth of plants. This is followed by accounts of the supply of water and nutrients to plant roots.

7.2 Plant development and growth

Seed and seedlings

The seed is the source of genetic material from which the plant develops. It consists of an embryo and a store of food reserves including carbohydrates, lipids, proteins and inorganic nutrients, enclosed in a protective coat. Its water content may be as low as 5% by mass. It is unable to germinate until it has passed through a period of dormancy, which may last from a few days to several years, depending on species. After this period seeds normally germinate when appropriate conditions are available, these being supplies of water and oxygen and a suitable temperature that allows the various metabolic processes in the seed to take place at an adequate rate.

In soil, the seed absorbs water if the water potential in soil (ψ_{soil}) is greater than that in the seed (ψ_{seed}). The components of ψ_{soil} that determine uptake are the osmotic potential and the capillary (matric) potential, as referred to in Sections 6.2 and 6.3. If the soil is only slightly moist or is salty, germination might not occur or will be slow. At the same time as water is absorbed, seeds of most plant species take up oxygen and release carbon dioxide. Oxidation and hydrolysis of the organic compounds in the seed provide the nutrients and energy requirement of the developing embryo. Successful germination of seeds in soil, at least for most plants, therefore depends on the soil being damp, and having the tilth and degree of compaction to allow them to absorb both water and oxygen.

In most plant species (there are several exceptions) the radicle (root) emerges first, anchoring the seed in the soil. This is followed by emergence of the shoot, which is known either as the coleoptile, the first leaf of grasses and related species, or as the plumule, which carries the primary bud in other species. Until the cotyledons or the first true leaves are able to photosynthesize, the seedling depends on the organic compounds in the seed. Depth in the soil is therefore crucial. If the seed is too near the surface the soil might become dry before it germinates, and if it is too deep the food reserves will be used up before the shoot emerges above the ground. Generally, the smaller the seed the shallower it should be in the soil.

Shoots

After the seedling has emerged above the surface of the soil the growing plant requires a suitable temperature, a supply of water and nutrients for uptake by the roots, and the interception of sunlight by the leaves. Further conditions, daylength and exposure to low temperatures, affect the development of some plant species and particularly the onset of flowering.

Some of the sunlight intercepted by green leaves is used in photosynthesis. Of the total radiation energy which reaches a leaf only about 1–5% is used for plant growth, most of the remainder being lost as sensible heat and latent heat of vaporization. An example of a radiation balance is shown in Table 7.1. For photosynthesis, plants obtain the required energy from radiation with wavelengths between 400 and 700 nm (in the

Table 7.1. *An example of daily radiation balance on a clear summer day*

	Energy $(MJ\ m^{-2}\ d^{-1})$
Income	
Incoming energy at surface	22.0
Reflected	4.5
Outgoing radiation	7.3
Net radiation at surface	10.2
Use at surface	
Evaporation	8.7
Heating air and plants	0.8
Heating soil	0.5
Plant growth	0.2
	10.2

Source: From Milthorpe, F.L. and Moorby, J. 1979 *An Introduction to Crop Physiology*, 2nd edition. Cambridge University Press; with permission.

wavelength range of visible light), the process requiring carbon dioxide and water. The overall reaction is

$$nCO_2 + nH_2O + energy \underset{\text{respiration}}{\overset{\text{photosynthesis}}{\rightleftharpoons}} (CH_2O)_n + nO_2. \qquad (7.1)$$

The biochemical reactions that lead to the production of carbohydrate, $(CH_2O)_n$, are fairly well understood but are outside the scope of this discussion. It needs to be mentioned, however, that plants use one of two pathways in the first stages of the assimilation of carbon dioxide.

In those known as C_3 plants, CO_2 is fixed by an enzymic reaction with ribulose diphosphate (a pentose phosphate) to produce phosphogly-ceraldehyde (a triose sugar); further enzymic steps then lead to the production of hexose sugars. This pathway, known as the reductive pentose phosphate (RPP) or Calvin cycle, is used by most of the temperate grasses, cereals, legumes and trees (Table 7.2). In some other species, known as C_4 plants, there is an additional pathway whereby CO_2 is first converted to oxaloacetic acid, which is rapidly converted to malic and aspartic acids. Subsequently these products are decarboxylated to produce CO_2, which is metabolized in the RPP cycle.

There are also anatomical differences between the two groups, enabling the C_4 plants to maintain a higher concentration of carbon dioxide in some of their chloroplasts than would result by diffusion from the air. This gives them a potential growth rate greater than that of C_3 plants. It also means that an increased concentration of carbon dioxide in the atmosphere has little effect on the growth of C_4 plants but increases the potential growth of C_3 plants, a point referred to in Section 11.11 on the effects of global warming.

Photosynthesis results in the production of sugars and other carbohydrates, which are used to provide energy, maintain the structural organization of plant cells and produce new leaves, shoots, roots and reproductive tissues. The energy required for these processes is obtained by respiration (Equation 7.1). In its effect, respiration is the

Table 7.2. *The main food crops of the world classified according to their use of a C_3 or C_4 pathway in the assimilation of carbon dioxide*

C_3 crop plants	wheat, rice, barley, potato, cassava, soybean, oats, banana.
C_4 crop plants	maize, sorghum, sugar cane, millets.

reverse of photosynthesis: oxygen is taken up and carbon dioxide is released, although the biochemical reactions are quite different. It occurs continuously, day and night.

The carbon available for plant growth is the difference between photosynthesis and respiration:

$$P_N = P_G - R, \qquad (7.2)$$

where P_G is total (gross) photosynthesis, R is respiration and P_N is net photosynthesis, also known as net primary production.

If there is sufficient light, water, nutrients and a suitable temperature a seedling in bare soil will grow for some time at an exponential rate, because photosynthesis of the plant increases as more leaves are produced. The rate becomes more nearly constant when there is self-shading, competition from other plants, or the plant stops producing new stems and leaves. A further stage is a decreasing growth rate as the plant approaches maturity; the rate becomes zero and then negative as the plant dies. When approaching maturity the plant transfers organic compounds and inorganic elements to reproductive and/or storage organs such as fruit, seed and roots. These stages are illustrated in Figure 7.1.

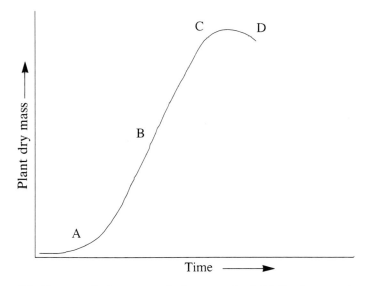

Figure 7.1 Change of plant mass during growth. Initially the growth rate is exponential (A), it becomes linear (B), decreases (C) and finally becomes negative (D) as the plant dies.

If M_2 and M_1 are the plant's dry masses at time t_2 and t_1 the average growth rate is

$$(M_2 - M_1)/(t_2 - t_1) \simeq dM/dt$$

for a short period.

For comparison of different plant species, or the same plant at different stages of growth, it is more instructive to use the relative growth rate, $(1/M)(dM/dt)$, which takes into account plant mass, M. Over a finite period of time, $(t_2 - t_1)$ the relative growth rate, R_M is given by the difference in plant masses expressed as their natural logarithms:

$$R_M = (\ln M_2 - \ln M_1)/(t_2 - t_1). \tag{7.3}$$

Roots

As mentioned above, the first root (radicle) usually emerges from the seed before the shoot. Dicotyledons usually produce only one root from the plant embryo whereas monocotyledons produce between one and about ten, depending on the species, and further roots from the nodes at the base of the leaf sheaths. These main roots are called axes; those that develop from the embryo are seminal axes and those from the nodes at the base of the stem are nodal axes. Lateral branches arise from the axes of nearly all plants: primary or first-order laterals from axes, secondary or second-order laterals from the primaries and so on (Figure 7.2).

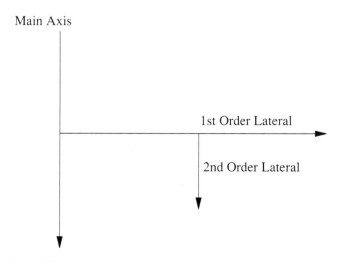

Figure 7.2 Schematic representation of root systems of plants.

Table 7.3. *The approximate sizes of roots of barley and wheat*

	Main axis	1st order lateral	2nd order lateral	Root hair
Diameter (cm)	0.05	0.02	0.01	0.001
No. per cm of root of next higher order	—	2	1	1000
Length (cm) per cm^3 of soil	1	5	2	1000

A root grows longer as a result of the division and elongation of cells near the root tip. The tip is surrounded by a sheath of mucigel (polysaccharide gel produced by cells at the root tip) which may act as a lubricant for root elongation through soil and has other effects discussed in Section 7.6. Extension of root axes is at a rate of 1–20 mm per day, or even higher, but growth of the laterals is slower. Root hairs emerge about 5–10 mm behind the root tip and may extend laterally a distance of about 1 mm. Table 7.3 gives measurements of root systems of barley and wheat.

Because roots provide the plant with water and nutrients, the depth of the rooting system and the volume of soil that it occupies can have a big effect on plant growth.

Under optimum conditions depth of rooting differs considerably between plant species. A plant that grows from seed to maturity in about 25 days may root to a depth of only about 50 mm. Wheat and barley roots can grow to a depth of 1–2 m and those of lucerne (*Medicago sativa*) to about 3 m. Roots of perennial grasses can also grow to these depths, and roots of some tree species can grow deeper. Usually, however, physical or chemical conditions in the soil restrict root growth, as discussed in the next section.

A plant with a deep root system has access to more water than one with shallow roots and may therefore survive a period of drought, but it is the roots near the soil surface that have access to the highest concentration of nutrients. The proportion of the total root system is usually greatest in the top 10–15 cm of soil, where the length of root per unit volume of soil (units cm cm^{-3}) varies between about 1 for some legumes to about 20–25 in grassland. It has been estimated that near the soil surface roots seldom occupy more than 5% of the soil volume and with more sparsely rooting species the value is less than 1%.

The growth of roots depends on the transfer of organic compounds from leaves, shoots and other photosynthesizing organs. During a season of growth about one third of the assimilate is transferred to the roots of annual species for growth and metabolism; perennial plants may transfer a higher fraction of up to about one half. Part of the carbon in the organic compounds transferred to roots is lost as CO_2 by respiration, exudation from roots, and oxidation by microorganisms as the roots decay. The carbon that remains in the roots of annuals gives a ratio of roots to shoots on a dry mass basis of $0.3 - 0.4$ during early growth, which decreases to about 0.1 at maturity as assimilate is transferred to the developing seed. With mature trees the ratio of roots to tops is usually about 0.2, but can be higher, and in permanent grassland the ratio is usually between 5 and 10 (see Table 5.2).

7.3 Restrictions to root growth

There are several soil conditions that restrict root growth:

Physical
Mechanical impedance, usually associated with high bulk density as occurs in compact layers.
Absence of cracks.
Shortage of oxygen, usually due to waterlogging.
Dry soil.
Temperature too low or too high.
Chemical
High aluminium concentrations, usually associated with low pH.
Low nutrient supply.
Phytotoxic chemicals in anaerobic soil.

Mechanical impedance and cracks

Mechanical impedance is one of the most common limitations to root growth. Roots will grow through continuous pores with a diameter bigger than themselves (about 300–500 μm), but as there are usually not enough big pores roots must also displace soil particles. They can only do this if the pressure exerted by the elongating cells at the root tip can overcome the resistance of the soil matrix to compression. This resistance depends on the composition of the soil and its bulk density, and becomes greater as the soil dries. The resistance can be measured by pushing into the soil a suitable probe attached to a device to measure force, known as a penetrometer (Figure 7.3). Bulk density is also

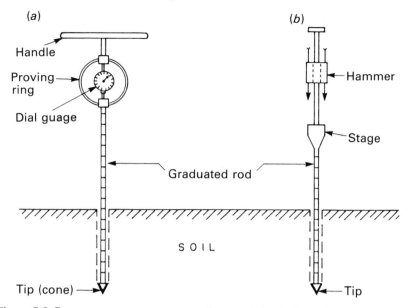

Figure 7.3 Penetrometers to measure soil strength in the field: (*a*) push-type and (*b*) hammer-driven, impact-type. (From Hillel, D., 1980. *Fundamentals of Soil Physics*, Academic Press, New York; with permission.)

sometimes used to indicate the conditions under which root growth will be impeded. Values of $1.55 - 1.85$ g cm^{-3}, depending on particle size distribution, may reduce root growth. These values are only a rough guide to the restriction of root growth, first because of the effect of soil water content on soil compressibility, and secondly because roots may grow through cracks and pores even in soils of high bulk density.

One of the purposes of soil cultivation is to reduce mechanical impedance brought about by raindrop impact, by the passage of machinery and by the feet of animals, which can compress the top 10–20 cm of soil and restrict root growth. At greater depths cracks are often present in soils containing 2 : 1 clays (see Section 2.7), which swell when wet and shrink when dry. Pores created by earthworms or roots from previous plants also allow roots to grow through the soil. Where none of these channels exist cracks may be created by subsoiling under suitable conditions (Section 8.2).

Aeration

The roots of all plants require oxygen for respiration. Many plants that grow in wetlands, and these include several species of bog plants and rice (*Oryza sativa*), transport oxygen internally to, and through, roots via air-filled cavities called aerenchyma. Some of the oxygen which passes into the roots may leak into the soil, creating oxidizing conditions in the immediate vicinity of the root in what is otherwise a reducing medium. Most plant species require oxygen from air-filled pore spaces in the soil, although it has been shown that some can develop aerenchyma if the soil becomes anaerobic. The stimulus for the production of aerenchyma appears to be the presence of ethylene.

The concentration of oxygen in soil air is less than that in the atmosphere for various reasons (Section 2.6), but it does not usually limit the supply to roots unless it falls to a low value, perhaps less than 10% by volume of the air. It is the volume and distribution of soil air that are critical. If soil becomes waterlogged the volume of air decreases towards zero. If there is partial waterlogging the diffusion of oxygen to roots is reduced because of the smaller area and greater tortuosity of the conducting pore space. Under these conditions the soil may become wholly or partly anaerobic.

Plants vary in their response to anaerobic conditions, shoots as well as roots being affected in sensitive species. As a generalization, root elongation is reduced when the gas phase to which they are exposed contains less than 10% of oxygen, which occurs only rarely. High concentrations of carbon dioxide may also reduce root elongation. In addition to low oxygen and high carbon dioxide concentrations there are other effects of anaerobic conditions on roots and shoots.

Several low-molecular-mass organic compounds that are produced in poorly aerated or waterlogged conditions affect root growth. The gas ethylene, C_2H_4, is produced by roots, and in poorly aerated soils also by microorganisms. At low concentrations it may stimulate root growth, but it has also been found to restrict growth in some species at 2 ppmv of the gas phase, a concentration which can be exceeded in anaerobic soils. Under anaerobic conditions carboxylic acids can be produced at concentrations that restrict root growth. They include acetic and butyric acids. Phenolic acids produced during the decomposition of plant material containing lignin are more persistent in anaerobic conditions and they also restrict root growth.

To summarize, root growth and the development of mesophytic (non-wetland) species is restricted by anaerobic conditions in waterlogged soils. There are several possible causes, which may differ in importance between soils and between plant species.

Effects of temperature and soil water content

In addition to the effect of excess water described above, the growth of roots is restricted in dry soils, although less than the growth of shoots. As soils become dry, root growth may be restricted more by the stronger cohesion of the soil particles than by low water potentials.

Temperatures outside the optimum range for a particular species reduce the growth of both roots and shoots. At a low temperature shoot growth is reduced more than root growth so the ratio of root to shoot increases.

Nutrient supply

A deficiency of any one of the essential nutrients reduces the mass of roots and shoots produced by a plant during its growth to maturity. The growth of roots is affected less than the growth of shoots. As a result, the ratio of root to shoot is increased by a low supply of nutrients.

The effect of nutrients, at least for phosphate and nitrogen (as NO_3^- or NH_4^+) is, however, different when placed in a localized zone in the soil. The high local concentration stimulates the initiation and extension of primary and secondary laterals. The mass of laterals in the zone of high nutrient status takes up a high proportion of the nitrogen and phosphate required for plant growth. This is one of the reasons why fertilizer is often placed below the seed.

Aluminium

Because of the coordination of growth between shoots and roots any element which is toxic to plants will reduce the growth of roots. Aluminium has, however, a direct effect on roots and in acid soils is a common cause of reduced plant growth. Aluminium is taken into the root where it temporarily inhibits cell division in the meristems of root tips and therefore stops elongation. The roots become stubby and brown. Because of the lack of elongation, roots are restricted to the upper part of the soil and the plants become subject to drought. Species and genotypes differ considerably in their sensitivity to aluminium.

7.4 Requirement of plants for water

The leaves of plants exist in a relatively dry environment compared with that surrounding their roots. The stomata on their surfaces, which are the channels for the exchange of carbon dioxide and oxygen during photosynthesis and respiration, also allow water vapour to escape into the atmosphere. Most of the water that escapes passes through the stomata although about 10% diffuses through the leaf cuticle. Evaporation of water from plant leaves serves to keep them cool and is known as transpiration. The meteorological factors that determine the rate of transpiration from a plant canopy are referred to in Section 6.6.

During vegetative growth the fresh mass of plants is 70–90% water, say 20 t ha^{-1}, whereas the daily transpiration loss during good growing conditions is twice or three times this, that is, about 50 t ha^{-1}. The proportion of water that is taken up and retained in plants is less than 1%, but is nevertheless essential for all the biochemical processes that take place in plant tissues.

Rainfall

The important characteristics of rainfall that affect the supply of water to plants are listed in Table 7.4. Where water supply limits plant growth, the distribution of natural species and the range of crops that can be grown depend largely on (i) annual rainfall, (ii) the length of the rainy season, and (iii) the number of days between showers. Some of the rain will recharge the soil profile, some will be evaporated from the soil surface, or be transpired by plants, some might run off the soil surface, and some might be discharged into rivers or drain into the groundwater (Figure 7.4).

Table 7.4. *Factors that affect the supply of water to plants*

Soil properties	Characteristics of rainfall
1. Depth of soil profile	1. Annual rainfall
2. Available water between permanent wilting point and field capacity	2. Length of rainy season
	3. Number of days between showers
3. Water infiltration rate	4. Intensity of storms
4. Rooting depth	5. Variability between years
	6. Season of the year in relation to the growing season (winter or summer rainfall)

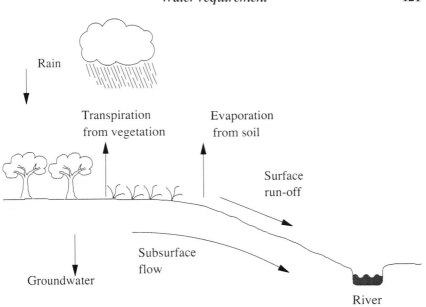

Figure 7.4 Pathways of water flow.

Storage of water in soil

The capacity of the soil to store water which can be extracted by plants affects both the length of the growing season and their survival during periods of several days without rain. Water storage in two deep soils at Hyderabad, India, is shown in Figure 7.5. The extractable water in the topsoil was assumed to be that held between field capacity and the permanent wilting point (Section 2.4), and at greater depths extraction was limited by low root densities. The general point to be made is that texture affects water storage in soil; in this comparison storage in the coarse-textured Alfisol was about 60% of that in the fine-textured Vertisol. The amount of extractable water that is stored also depends on the depth of the soil profile, the structure of the soil and the depth of rooting.

Infiltration of water

If the rate of rainfall is greater than the infiltration rate into the soil there is surface run-off, which can cause erosion (Chapter 12). The loss of water can be serious where it is in short supply, although it might be useful where it collects in valley bottoms. Soils with high contents of fine

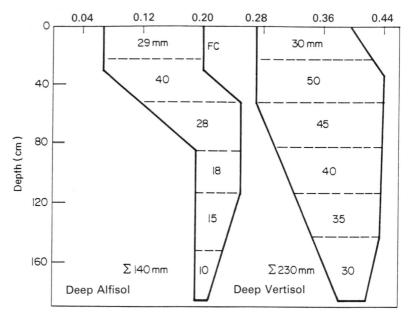

Figure 7.5 Amounts of water that can be extracted by crops growing in two soils in India. (From Russell, M.B. 1980. In *Proceedings Agroclimatological Research Needs of the Semi-Arid Tropics*, pp. 75–87. ICRISAT, India; with permission.)

sand and silt, which usually have weak structure, have low infiltration rates, as do soils that have been compressed by the passage of heavy implements or have been trampled by animals or man.

Rate of supply of water, in addition to the amount of available water in the root zone, is also important. As soil becomes drier the rate of movement of water within the soil rapidly decreases (Chapter 6) and the supply to the roots becomes less than the loss by the leaves. Unless there is rain or the plants are irrigated, the movement of water becomes too slow for the leaves to recover at night and the plants become permanently wilted. The practice of irrigation is discussed in Section 8.10.

Survival of water shortage by plants

Plant species that are able to survive periods of water shortage may have any of the following characteristics:

(i)   a short life span which allows them to grow from seed to maturity in a few days;

(ii)   a deep and widely distributed root system for absorbing water;

(iii)   ability to store water in above-ground or below-ground tissues;

(iv)   a mechanism for reducing transpiration loss, including the closure of stomata, turning of leaves away from the sun and shedding leaves.

Depending on the retention and availability of water in the soil, some plant species need rain every few days during growth if they are to survive; others can survive several months without rain.

7.5 Requirement of plants for nutrients

The elements required by higher plants are listed in Table 7.5. Conventionally they are grouped according to the amounts taken up by plants into macronutrients and micronutrients. Nitrogen and potassium are usually taken up in the largest amounts, which can exceed 100 kg $ha^{-1}a^{-1}$, whereas the uptake of molybdenum at the other extreme can be less than 1 g $ha^{-1}a^{-1}$. The relative amounts of elements taken up from soil are shown in Table 7.6.

Requirement for the elements in Table 7.5 has been demonstrated by growing plants in solutions from which the element under investigation has, as far as possible, been removed. The biochemical function of each essential element is known. The list is believed to apply to all species whether or not a requirement by the species has been demonstrated.

The elements are taken up by plant roots from the soil solution. Nitrogen is taken up as NH_4^+ and NO_3^-, phosphorus as $H_2PO_4^-$ and HPO_4^{2-}, sulphur as SO_4^{2-}, and the metals as cations, for example as

Table 7.5. *Elements required by higher plants*

From the atmosphere and water	carbon, hydrogen, oxygen
From the soil:	
Macronutrients	nitrogen, phosphorus, potassium, calcium, magnesium, sulphur
Micronutrients	iron, manganese, copper, zinc, boron, molybdenum, chlorine, nickel
Beneficial elements	cobalt[a], sodium and silicon

[a]Cobalt is essential for biological nitrogen fixation by bacteria, including those species that have a symbiotic relationship with plants.

Note that nickel is included in this list. It is the most recent addition and appears to meet the required criteria for essentiality.

Table 7.6. *Concentration in plants of nutrient elements absorbed from soil that are considered adequate*

Element	Concentration in dry matter (mg kg^{-1})	Relative number of atoms with respect to molybdenum
Molybdenum	0.1	1
Copper	6	100
Zinc	20	300
Manganese	50	1000
Iron	100	2000
Boron	20	2000
Chlorine	100	3000
	(%)	
Sulphur	0.1	30 000
Phosphorus	0.2	60 000
Magnesium	0.2	80 000
Calcium	0.5	125 000
Potassium	1.0	250 000
Nitrogen	1.5	1 000 000

Source: From Epstein, E. 1972. *Mineral Nutrition of Plants: Principles and Perspectives.* Wiley, New York; copyright © 1972, reprinted by permission of John Wiley & Sons, Inc.

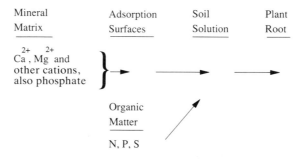

Figure 7.6 Sources of nutrient elements in soil and their transfer to plant roots.

Ca^{2+}, Mg^{2+} and K^+. The nutrient elements can be carried to the root in the soil solution, which moves to the root at a rate determined by transpiration: this process is known as mass flow. Nutrients can also reach the root surface by diffusion. These two processes are described in Section 6.8. Soil microorganisms also influence the absorption of nutrients (Sections 7.6 and 7.7).

The soil solution is replenished by desorption of the nutrient elements held on the surfaces of clays and organic matter (Figure 7.6). Replenishment of the soil solution with nitrogen, phosphorus and sulphur also occurs as a result of the mineralization of soil organic matter.

Soils receive nutrients by rock weathering, biological fixation of nitrogen, deposition from the atmosphere, mineralization of leaf litter and, in agricultural systems, in fertilizers and manures. The additions of nutrients and their cycling in natural ecosystems will be considered in Section 7.8. Agricultural systems are dealt with in the next chapter (Section 8.3).

7.6 The rhizosphere

The rhizosphere is that part of the soil whose properties are influenced by the presence of roots. The dimensions cannot be given with any precision. Most of the effects occur within 1 or 2 mm of the root surface, but if the roots have mycorrhizas (Section 7.7) the effect can extend a few centimetres into the soil.

The effects of roots on soil properties arise from five causes.

1. Respiration by roots of most plants uses up oxygen in the soil air and releases carbon dioxide; as noted in Section 7.2, however, plants with well developed aerenchyma leak oxygen from their roots into the rhizosphere.

2. Absorption by the root of water and nutrients creates gradients of water potentials and nutrient concentrations close to the root surface (see Sections 6.6 and 6.8).

3. The ratio of cations to anions taken up by the root influences the rhizosphere pH. In order to maintain a balance of charge within the plant either protons or bicarbonate ions are excreted:

if uptake of $C_c > A_c$, protons are excreted, and pH of the rhizosphere decreases;
if uptake of $C_c < A_c$, bicarbonate ions are excreted and pH of the rhizosphere increases,

where C_c and A_c are the total cation and anion charge respectively. Thus, to balance the uptake of one calcium ion (Ca^{2+}), two nitrate ions (NO_3^-) are needed. An imbalance of uptake is one cause of soil acidification (Section 9.5).

4. Roots of most plant species develop an association with fungi

(mycorrhizas) whose hyphae penetrate the root as discussed in the next section.

5. Organic substances released from roots stimulate the growth of microorganisms and may also directly affect the uptake by roots of some nutrients including iron and manganese.

Roots therefore affect the properties of their own environment so that some of the properties of the rhizosphere, for example pH, composition of the soil air and microbial activity, are different from those of the bulk soil. There are consequential effects on the mineral nutrition of plants.

Of the five influences on rhizosphere properties listed above only the release of organic substances will be discussed here.

The three sources of organic substances are:

1. Mucilage (mucigel), a polysaccharide composed of hexose and pentose sugars and uronic acids, which is produced at the root tip. It extends a short distance along the root before being decomposed by microorganisms.
2. Low-molecular mass substances that include sugars, organic acids and amino acids which are excreted by root hairs and other intact cells.
3. Cellular material released by senescence of the root epidermis and cortex, and root hairs.

The proportion of plant carbon released into the rhizosphere as organic substances has been estimated or measured by several workers. Values within the range 2–20% have been reported, depending on the species and age of the plant. The release is higher in plants that have experienced stress during growth.

The organic substances provide a source of energy and nutrients for microorganisms. Numbers of bacteria per unit volume may be between 5 and 100 times higher in the rhizosphere than in bulk soil, and fungi, protozoa and nematodes may also be more numerous. The exudates and their partial decomposition products may form chelates with metals such as copper, and organic acids may release phosphate into solution, the effect being to increase the availability of the nutrients to plants. The rhizosphere microorganisms might consume nutrients that would otherwise be available to plants, but evidence for a significant effect is not convincing. Denitrification may be more rapid in the rhizosphere because of the reducing conditions induced by the greater respiration of the microorganisms.

There are experimental difficulties in investigating rhizosphere soil. It is not a sharply defined entity, and each property changes with distance from the root surface until it becomes the same as in the bulk soil; also, there is no reason for believing that the outer boundary is the same for all properties. At the inner boundary of the rhizosphere, whether at the root surface or in the decomposing cells of the root cortex, the rhizosphere affects plants in two ways. First, as mentioned above, it influences the uptake of some mineral nutrients. Secondly, populations of microorganisms responsible for root diseases may increase in the rhizosphere or be controlled by other organisms. Symbiotic fixation of nitrogen in nodulated legumes and some non-legumes also occurs at the inner boundary and is discussed in Section 5.5.

7.7 Mycorrhizas

Most plant species have a symbiotic association of their roots with specialized fungi. The associations are of mutual benefit and are known as mycorrhizas, meaning 'fungus roots'. The fungus benefits by receiving carbon compounds from the plant, and the fungus supplies the plant with mineral nutrients, of which phosphate is generally the most important.

There are two common types of association. Ectomycorrhizas form a sheath or mantle of fungal mycelium over the surface of fine roots (Figure 7.7). The hyphae extend outwards into the soil and inwards into the intercellular spaces of the root cortex to form the so-called Hartig net. This type of mycorrhizal association occurs commonly on the roots of trees, there being some specificity between tree and fungus species. Most of the fungi are basidiomycetes or ascomycetes. Because of the presence of the fungal mantle on fine roots, the presence of this type of mycorrhiza can often be recognized with the aid of a magnifying glass.

The term 'endomycorrhiza', which describes the other main group, has been largely replaced by the term 'vesicular–arbuscular mycorrhiza' (VAM), though the two are not synonymous. The term VAM describes the formation by fungi of the family Endogenaceae of structures called vesicles and arbuscules within the cells of the root cortex (Figure 7.8). They appear to serve as organs for the storage and transfer of carbon compounds and mineral nutrients between the fungal hyphae and the host plant. Infection by VAM has been observed in a wide range of plants. Exceptions are members of the Cruciferae and Chenopodiaceae.

Other types of mycorrhizal association are known. One carried by ericaceous plants has characteristics of both VAM and ectomycorrhizas.

Soil and plant growth

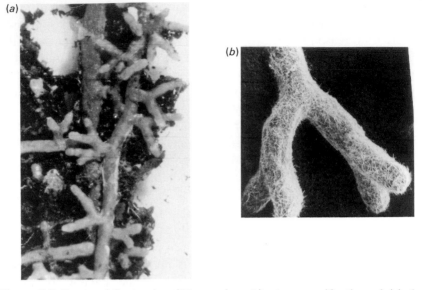

Figure 7.7 Roots of Scots pine (*Pinus sylvestris*) at a magnification of (*a*) 6 times, and (*b*) 30 times. The individual hyphae can be seen at the higher rate of magnification. Ectomycorrhizas are found on the roots of many tree species. (Photographs by Dr D. Jones, The Macaulay Land Use Research Institute, Aberdeen.)

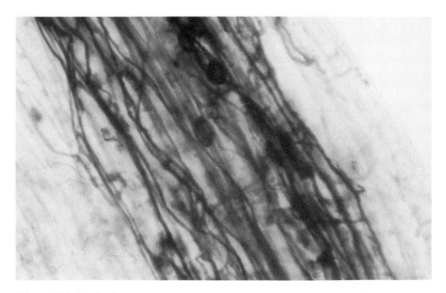

Figure 7.8 Vesicular–arbuscular mycorrhiza in an onion root after staining with trypan blue. Several vesicles can be seen, one being clearly visible in the centre of the photograph; magnification × 100. The hyphae external to the root are not shown. (Photograph P.J. Harris.)

Table 7.7. *Effect of mycorrhizas on the growth, phosphorus concentration and phosphorus inflow into onions growing in soil*

	Mycorrhizal	Control
Total plant fresh mass (g)	13.05	2.22
P (% in dry matter)	0.31	0.14
Mean inflow into roots (mol m^{-1} s^{-1})	17×10^{-12}	3.6×10^{-12}

Source: From Nye, P.H. and Tinker, P.B. 1977. *Solute Movement in the Soil-Root System.* Blackwell Scientific Publications, Oxford; with permission.

Mycorrhizas are very important in nutrient uptake by plants, especially in soils low in nutrients. Increased uptake of P, N, K, Cu, Zn and Ca has been demonstrated. Experiments are usually conducted by growing plants in sterilized soil (to kill the native microorganisms) and comparing growth in unamended soil with that in soil to which fungal spores have been added. Table 7.7 shows that with this kind of comparison, the inflow of phosphorus was about four times as fast into the infected roots as in the controls. When the labile (readily available) P has been uniformly labelled with ^{32}P, the ratio of ^{32}P to ^{31}P in mycorrhizal roots has been found to be the same as in non-mycorrhizal roots. This observation indicates that the same pool of labile P is used by the plants in both treatments; that is, the hyphae of the VAM do not make soluble any non-labile P. The hyphae give the plant roots access to a greater volume of soil from which nutrients can be absorbed and transported directly to the root. This is particularly beneficial for the absorption of ions such as phosphate which has low mobility in soil. In effect, the hyphae allow the root to reach a bigger pool of phosphate, and in soil low in phosphate this can increase plant growth. It has also been shown that carbon compounds, water, phosphate and probably other nutrients can be transferred via the hyphae between mycorrhizal plants, suggesting a degree of interdependence of plants in natural communities.

Although inoculation of sterilized soil by mycorrhizal fungi has a beneficial effect on the growth of plants under conditions of nutrient deficiency, there has been very little success with arable crops on a field scale on soils which have not been sterilized. Mycorrhizal infection does, however, explain how the heavily mycorrhizal roots of cassava (*Manihot esculenta*) enable the crop to grow well in phosphate-deficient soil whereas other crops fail. The success of species of wild plants on soils low in nutrients probably has the same explanation. Several tree species have been found to respond to inoculation when introduced into

a new area. Inoculation is often done by using soil from forests or tree
nurseries where the same species has been grown previously; inocula of
the required fungus are also used.

7.8 Nutrients in natural ecosystems

The nutrients taken up by plants from soil originate under natural conditions from rock weathering and the atmosphere (Figure 7.9). The atmospheric sources from which most of the soil nitrogen and sulphur are
derived include precipitation (rainfall and snow), several gases (nitrogen, ammonia, nitrogen oxides, sulphur oxides etc.) and aerosols.
Weathering of rock minerals releases the nutrient elements calcium,
magnesium, phosphorus, potassium, sodium and the micronutrient
elements.

The relative importance of the atmosphere compared with mineral
weathering was shown by an investigation in New Hampshire, USA, in
the Hubbard Brook Experimental Forest. Forty-five years before
monitoring of the site began the forest had been felled and the standing

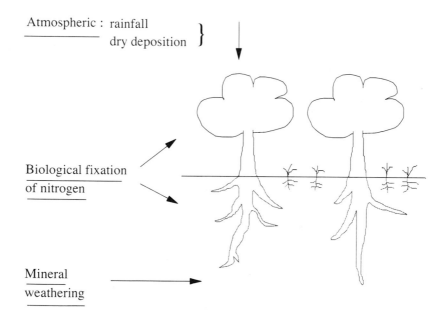

Figure 7.9 The three sources of nutrients for plants growing under natural
conditions. Note that biological fixation of nitrogen by bacteria occurs in the
nodules of plant roots, to a less extent in soil and probably occurs in the canopy
of trees.

mass of new trees was increasing annually. The nutrient inputs were either stored in the vegetation or lost by drainage as streamflow, the assumption being made that the composition of the soils (mostly spodosols) did not change over a 12-year period. The budgets of six elements are shown in Figure 7.10. Each input is represented as a percentage of the total input for each element; storage and loss are similarly represented as percentages of the total. The figure shows that rock weathering was the main source of calcium, magnesium, potassium and sodium, and that precipitation, gases and aerosols were the main source of nitrogen and sulphur. Elements in the first group have been termed 'non-volatile', and in the second group 'volatile' because they form volatile compounds at some stage in their cycles.

The stock of 'non-volatile' nutrient elements in soil depends on the composition of the minerals in the parent material, including dust from the atmosphere, and the extent of leaching. Because of the variability in composition of the parent material, the stock of 'non-volatile' elements in undisturbed soils differs more between sites than that of the 'volatile' elements, which are derived from the atmosphere.

The contrast between the calcium and nitrogen contents of soils under four tropical forests (Table 7.8) serves as an example. The Amazon Caatinga forest of Venezuela and the Heath forest of Sarawak grow on soils low in 'non-volatile' nutrients, shown by the low contents of exchangeable calcium. In the soils from the other two forests the contents are about 20 times as large. There is a sharp contrast between the two soils from Sarawak, which are 20 km apart. Their nitrogen contents are the same but their calcium contents are very different. At all four sites the nitrogen contents of the soils are more similar than the calcium contents. The data in Table 7.8 also show that the above-ground biomass of the forests is an unreliable guide to the nutrient content of the soils, a fact that is often overlooked when forests are cleared for agricultural development.

Additions of nutrients

The amounts of nutrients supplied to natural ecosystems in rainfall can be measured directly. The amounts depend on the rainfall and the proximity of the site to the sea, sources of atmospheric pollution and atmospheric dust. Tables 7.9 and 9.6 give the annual additions at relatively clean sites. They show that rainfall is an important source of nutrients when considered over a long period of time.

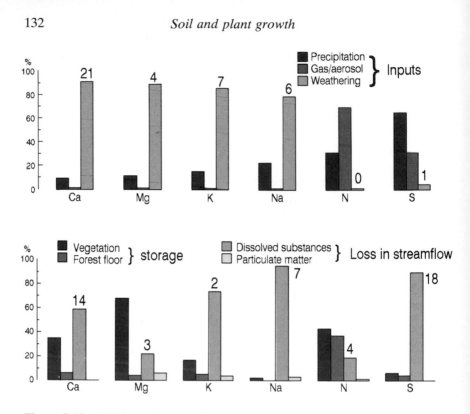

Figure 7.10 Additions, losses and storage of nutrients in the Hubbard Brook Experimental Forest expressed as percentages of their summation; numbers on the columns are kg ha^{-1} a^{-1}. Note that inputs of cations are mainly from mineral weathering whereas inputs of nitrogen and sulphur are mainly from the atmosphere. (From G.E. Likens *et al.*, 1977. *Biogeochemistry of a Forested Ecosystem*. Springer-Verlag, New York.)

More difficult to measure are the additions by dry deposition and aerosols, which supply mainly nitrogen and sulphur. They are discussed in Section 9.7 in relation to soil acidification. There can also be large additions of nitrogen by biological fixation of atmospheric nitrogen (Section 5.5). On land which is newly colonized by plants capable of fixing nitrogen, annual additions of between 40 and 130 kg N ha^{-1} have been measured.

The rate of nutrient release from the weathering of minerals in rocks cannot be measured directly. In the Hubbard Brook study referred to above the amounts released by weathering were calculated as (loss in stream water) plus (accumulation in biomass and litter) minus (addition

Table 7.8. *Nitrogen and calcium in the above-ground biomass of four old tropical forests, the underlying soil, and in annual litterfall*

	Caatinga forest, Venezuela (1)	40-year forest, Ghana (2)	Alluvial forest, Sarawak (3)	Heath forest, Sarawak (4)
Above-ground biomass (t ha^{-1})	280	336	250	470
Total N in biomass (kg ha^{-1})	334	1832	n.d.	n.d.
Total Ca in biomass (kg ha^{-1})	238	2527	n.d.	n.d.
Soil total N (kg ha^{-1})	786[a]	4596[b]	7800[c]	7800[c]
Soil exchangeable Ca (kg ha^{-1})	133[a]	2578[b]	1600[c]	62[c]
C in litterfall[d, e] (t ha^{-1} a^{-1})	2.5	4.7	5.2	4.1
N in litterfall[d] (kg ha^{-1} a^{-1})	42	200	111	55
Ca in litterfall[d] (kg ha^{-1} a^{-1})	31	206	286	83

[a] 0–40 cm
[b] 0–30 cm
[c] 0–30 cm
[d] fine litter
[e] assuming 45% C in litter dry matter.
n.d. – not determined
Sources: Forest (1), Herrera, R., PhD Thesis, Reading University; forest (2), Nye, P.H. and Greenland, D.J. 1960, (see also Table 7.9); forests (3) and (4), Proctor, J. *et al.* 1983, *Journal of Ecology* **71**, 261–83 and related papers.

in rainfall). The annual values are shown in Figure 7.10. The values at other sites can be expected to vary with the composition of the parent material and the rate of weathering. The nutrients that are released will be available for plant uptake if the release is within the root zone, but in

Table 7.9. *Additions of nutrients to the forest floor, Ghana*

	Nutrient elements (kg ha^{-1} a^{-1})				
	N	P	K	Ca	Mg
Timber and litter	235	10.2	74	288	53
Throughfall (rain wash from leaves)	12	3.7	220	29	18
Total	247	13.9	294	317	71
Rainfall (in open)	15	0.4	18	12	11

Source: From Nye, P.H. and Greenland, D.J. 1960. *The Soil under Shifting Cultivation.* Technical Communication No. 51, Commonwealth Bureau of Soils, Harpenden, UK.

deeply weathered tropical soils the nutrients may pass directly into the drainage water.

Nutrient cycling

In natural ecosystems there are additions and losses of nutrients, and accumulation in biomass, as described above; there is also internal cycling (Figure 7.11).

Nutrients taken up by plants are deposited on the soil surface in litter (leaves, flowers, fruits, twigs, branches, fallen trees), throughfall, also known as leaf wash, and in stem flow. Amounts of carbon and some mineral nutrients in the litterfall of some tropical forests are given in Tables 7.8 and 7.9.

Large amounts of potassium, and usually also of chloride, are washed from leaves together with smaller amounts of phosphorus, calcium, magnesium and sulphate but only small amounts of nitrogen. In the example in Table 7.9 the nutrient content of rain increased on passage through the foliage of the trees. Leaves of trees with very low nutrient concentrations may, however, remove nutrients from rainwater.

The cycling of nutrients conserves them against loss by leaching and volatilization. Another mechanism for nutrient conservation in trees is the withdrawal of nutrients from maturing leaves into woody tissues. Nevertheless, there are leaching losses at sites with through-drainage, and if the supply of nutrients from rock weathering is less than the loss by leaching the soil stock of nutrients becomes depleted. Many soils of the world under natural vegetation have low levels of nutrient elements

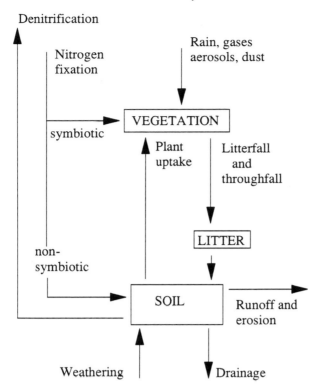

Figure 7.11 The nutrient cycle.

so that remedial measures are needed when the land is converted to agricultural use.

7.9 Summary

Plant growth and development are strongly influenced by soil conditions, mainly because the physical and chemical conditions in the soil affect the development of the plant root system. The distribution of roots is important because roots take up water and most of the nutrients, and provide anchorage.

We have still much to learn about plant roots. They affect the properties of soil in their immediate vicinity (the rhizosphere) in a variety of ways that are not yet properly understood. Mycorrhizal associations can increase nutrient uptake and occur on the roots of most species in natural communities; more needs to be known about them.

In natural ecosystems, rainfall provides water. Soil nutrients originate in mineral weathering, deposition from the atmosphere and the biological fixation of nitrogen. Cycling of nutrients conserves them, but depletion occurs by leaching, erosion, and by gaseous loss of nitrogen and its compounds.

8
Soil conditions and crop production

8.1 Introduction

Management of an enterprise in agriculture, horticulture and forestry has to take into account a wide range of issues: the management of soils, crops and animals, the selection and use of machinery and implements, marketing arrangements, man management, and local and world commodity prices. Except for subsistence farming, where security of food supplies is all-important, the purpose is to make the enterprise profitable.

The objective of soil management is to create suitable conditions for the crops that are to be grown. As listed in Section 7.1, soil is required to provide anchorage and the physical and chemical conditions required by the plants. What the farmer does to help meet these requirements depends on what crops he grows, the required yield, the inputs that are available to him and the soil and climatic conditions.

Management of the soil started with the first farmers. Cultivations, rotations and irrigation are ancient practices. In the past 200 years there have been several innovations: farm machinery has become more powerful, crop varieties have been bred that give higher yields, and fertilizers and chemicals for the control of pests have been introduced.

Soil is the growers' main resource and it is in their interests to maintain it in as fertile a condition as possible. In this they are usually successful, but there have been, and still are, examples of bad management. Erosion has been caused by cutting down trees, salts have accumulated in soils under irrigation, and unsuitable soils have been brought into cultivation. Problems such as these, and ways of dealing with them, are dealt with in later chapters. Here we consider the current practices that are used to improve and maintain soil fertility; by soil fertility will be meant the capacity of the soil to produce the crops being grown.

8.2 Cultivations

The purpose of soil cultivations is to create suitable physical conditions for crop plants, namely:

1. To obtain a seed bed, that is, to create aggregates of a suitable size close to the soil surface in order to provide water and oxygen for the seed to germinate and allow the seedling to emerge above ground.
2. To provide conditions suitable for root growth by loosening the soil near the surface, and sometimes also at depth, in order to provide channels for the roots to grow, and to improve aeration (or drainage).
3. To bury weeds and crop residues, which would otherwise make it difficult to create a good seed bed; also to prevent weeds from competing with the crop.
4. There might be special requirements, for example to create furrows for planting a crop such as potatoes, or ridges which are used in many countries in preference to sowing a crop on the flat. Terraces to control erosion are another special requirement and are referred to in Section 12.8.

Under many conditions the requirements for creating a seed bed are met by first ploughing with a mouldboard plough, which inverts a slice or furrow to a depth of about 20–25 cm. The inverted soil is then reduced to a suitable tilth by using discs and harrows. If the bulk density of the soil surface is too low it is compressed by using a roller. Under

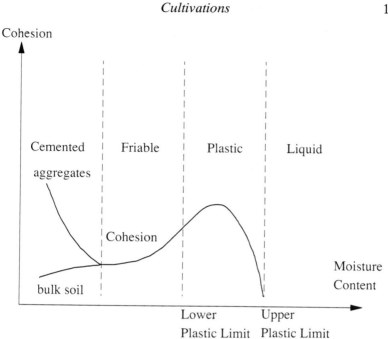

Figure 8.1 The four physical states of soil as it becomes increasingly wet. When plastic the soil sticks to farm implements and cultivations should be done when the soil is friable.

other conditions it is preferable to avoid inverting the soil. Tine blades or discs are used to make a suitable tilth without having to bury crop residues.

An experienced farmer can judge whether the soil is friable, that is, has the right water content for cultivation. If it is too wet the soil structure can be damaged: tractor wheels slip and the smearing of the soil surface blocks drainage channels, the ploughshare compresses the soil causing a plough pan, and the soil becomes moulded into clods, which might be difficult to break when they have dried. During cultivation the soil should be drier than the lower plastic limit, which restricts the number of days the soil can be cultivated. This is especially so if, after a soil has drained to field capacity, it remains wetter than the lower plastic limit (Figure 8.1). The soil should then be allowed to dry before it is cultivated. However, if it has a high clay content and is too dry, cultivation with discs and harrows may leave large aggregates intact making it difficult to convert the cloddy soil into a good seed bed.

Heavy machinery compresses the soil, especially if it is wet.

Compaction below wheels is most likely to occur during the mechanical harvesting of crops late in autumn, and the spreading of slurry in winter. When compression is shown to occur below the depth to which the soil can be ploughed, loosening can be brought about by drawing a deep tine behind a tractor when the soil is not too wet at depth, a technique known as subsoiling.

Drainage

After a period of rain, even if the soil has not been compressed, water might stand on the soil surface. There can be three main causes.

1. The soil overlies impermeable clay. This can be dealt with by installing drains that lead the water into ditches under gravity, and which in turn drain into a river system.
2. The soil is low-lying and there is a high water table. Drains will probably be required and water will need to be pumped out of the ditches into rivers.
3. The soil receives water from up-slope. This can be prevented by installing interception ditches or drains. Surface run-off from upslope can be reduced or prevented by contour ploughing and other techniques (Section 12.8).

A drainage system is also required where irrigation is practised in an environment where there is normally no through-drainage and surplus water has to be added to prevent the accumulation of salts (Section 8.10).

Direct drilling

Conventional cultivations are expensive and have the disadvantage that they leave the soil exposed to wind and rain and might therefore be subject to surface damage and erosion. An alternative is to drill seed direct into the stubble of a previous crop, a system known as 'no-till' or 'direct drilling'. Herbicides are used to control weeds. The system saves costs, but is not widely practised in the United Kingdom because many soils require cultivations to incorporate crop residues and produce a good seed bed. In other countries direct drilling is one of the techniques that is used to conserve soil from erosion.

Table 8.1. *Nutrient removal in harvested crops (kilograms per hectare)*

	Dry matter yield (t ha^{-1})	N	P	K	S	Ca	Mg
Wheat grain	5	100	20	28	8	3	8
Wheat straw	5	35	4	40	5	18	5
Maize grain	5	100	20	30	5	10	8
Maize straw	5	50	10	60	8	10	10
Rice grain	5	90	20	25	8	5	1
Rice straw	5	20	5	50	5	15	8
Ryegrass hay	10	160	30	180	12	40	12

Note: Values in the Table, which are from various sources, give only the order of magnitude because nutrient removal varies with crop variety and depends on the size of the nutrient supply.

8.3 Provision of nutrients

All agricultural systems involve the harvesting of crops or animals and the removal from the land of the nutrients they contain. Grazing animals remove less than crops, for although their intake can be high most of the nutrients are returned to the land in faeces and urine. The removal in crops depends on the total yield, the part of the crop that is removed in the harvest, and the particular crop that is grown. Table 8.1 gives some typical amounts. Nutrients are also lost in drainage water and some by volatilization.

If there is no replacement of nutrients then crop yields will decrease. Also, when land is first brought into cultivation there might be a need to correct nutrient deficiencies. These two problems have been overcome in four ways.

1. To lessen the rate of depletion nutrients are returned in crop residues, in organic manures from animals which are penned, and in some countries in night soil (human excreta) or sewage.
2. Nitrogen is added by biological nitrogen fixation from legumes or other biological systems (Section 5.5).
3. Nutrients are allowed to accumulate under a fallow, such as a bush fallow as has been traditionally used in the tropics.
4. Fertilizers are used to provide the nutrients that would otherwise be deficient and in the amounts required.

Table 8.2. *Composition of fertilizers in common use*

	N, P or K content (%)	Nutrient supplied as
Nitrogen fertilizers		
Ammonium sulphate	21	$(NH_4)_2SO_4$
Ammonium nitrate	32–34.5	NH_4NO_3
Calcium ammonium nitrate	25	NH_4NO_3 plus $CaCO_3$
Urea	46	$CO(NH_2)_2$
Anhydrous ammonia	82	NH_3 gas
Aqueous ammonia	27	NH_3 in water
Phosphorus fertilizers		
Single superphosphate	8–9	$Ca(H_2PO_4)_2$ plus $CaSO_4 \cdot 2H_2O$
Triple superphosphate	20	$Ca(H_2PO_4)_2$
Monoammonium phosphate	26	$(NH_4)H_2PO_4$
Diammonium phosphate	23	$(NH_4)_2HPO_4$
Rock phosphate	12.5–15.5	$Ca_5(PO_4)_3F$, but variable
Potassium fertilizers		
Potassium chloride	50	KCl
Potassium sulphate	42	K_2SO_4

8.4 Fertilizer use

The bulk of fertilizers now in use are manufactured by the chemical industry. Nitrogen, phosphorus and potassium are the nutrients supplied in greatest amount; others are included as needed. Table 8.2 lists the fertilizers in common use. Other sources of nutrients that are applied to crops are organic manures (Section 8.6), and, to a much smaller extent, processed parts of animals such as bone meal, fish meal, crushed hoof and horn, and dried blood. Chemicals which occur naturally, including sodium nitrate and kainite, are used in comparatively small amounts.

Three developments have been important in establishing the place of fertilizers in modern agriculture.

1. Long-term field experiments started in the 1840s at Rothamsted Experimental Station, Harpenden in southern England, showed that crop yields such as those of wheat could be maintained with continuous cropping when the required nutrients were applied (Table 8.3). In practice, however, good management usually also requires a crop rotation and often the addition of organic matter.

2. The synthesis of nitrogen fertilizers in the early 1920s from atmospheric nitrogen, based on the Haber process, made possible large-scale

Table 8.3. *Yield of wheat grain (tonnes per hectare) on Broadbalk field, Rothamsted Experimental Station*

Treatment[a]	1852–61	1902–11	1970–78
None	1.12	0.80	1.65
FYM	2.41	2.62	5.84
PKMg(Na)	1.29	1.00	1.78
NPKMg(Na)	2.52	2.76	5.32

[a]Treatments started in 1843. Annual rates of application were 144, 35 and 90 kg ha^{-1} for N, P, and K, respectively; FYM (farmyard manure) supplied 248, 43 and 325 kg ha^{-1} of N, P, and K, respectively.
Source: Rothamsted Experimental Station Report for 1982, Part 2, and earlier papers.

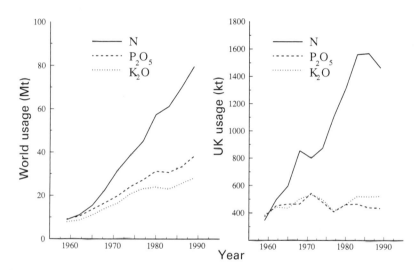

Figure 8.2 Annual use of fertilizers in the world and in the UK. (Derived from *FAO Yearbooks, 'Fertilizer'*. FAO, Rome.)

production at a low price. Later developments in the oil industry provided cheap energy for the process.

3. The introduction of higher-yielding, mainly short-stemmed, varieties of cereal crops has made it profitable to add greater amounts of fertilizer. Improved cultivations, irrigation and the use of pesticides have also justified the use of more fertilizer.

The use of fertilizer world-wide has increased dramatically since the 1950s (Figure 8.2). Fertilizer nitrogen has shown the biggest increase,

Table 8.4. *Causes of nutrient depletion in soil*

Cause	Nutrients most readily lost
Erosion	all nutrients
Surface run-off	NO_3^-, K^+, Ca^{2+}, Mg^{2+}, SO_4^{2-}
Leaching	NO_3^-, K^+, Ca^{2+}, Mg^{2+}, SO_4^{2-}
Crop removal	all nutrients
Gaseous loss	N, possibly S
Burning of vegetation	N, S

and it is projected that its use each year will reach 100 million tonnes by the year 2000. The need for more nutrients is to produce the food required by a rapidly increasing world population. In the United Kingdom the use of phosphorus and potassium levelled off after most of the initial shortages were corrected, and the need for nitrogen is unlikely to increase now that food production is meeting requirements.

The purpose of fertilizer use is to remove the limitation to crop growth that would be caused by an inadequate supply of nutrients in the soil. When land is first brought into cultivation there might be severe nutrient deficiencies, as of phosphorus in Australia and the South American cerrado. More commonly, nutrients that have been cycled through vegetation and have accumulated near the soil surface are sufficient for good crop yields for one or two or even several years, but nutrient depletion inevitably occurs (Table 8.4). The requirement might be for one or more of the 14 essential elements (Table 7.5), but in the United Kingdom and often elsewhere it is most commonly for nitrogen, phosphorus or potassium.

Cost/benefit of fertilizer use

With the main exception of subsistence farming, the products of agriculture are sold for cash. If the cost of an input such as fertilizer is to be justified, it must be less than the market value of the extra product given by the input. Increasingly, subsistence farmers need cash to purchase goods and for other purposes, and may pay for inputs in order to increase their saleable products.

The relation between cost of fertilizer and value of the product is shown schematically in Figure 8.3. On the steep part of the response curve each kilogram of fertilizer N commonly increases the yield of wheat grain by 25 kg. Given that the cost of fertilizer N is about £0.3 per

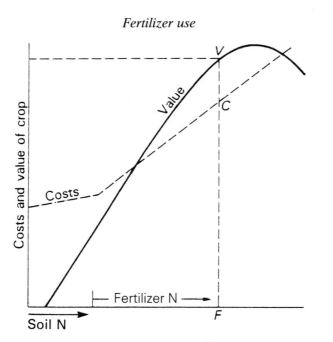

Figure 8.3 Profit from fertilizer use. Without fertilizer the value of the crop may not meet the fixed costs and excessive use reduces profits.

kg and the market value of the grain is about £0.1 per kg, it will be seen that the investment in fertilizer nitrogen can be very profitable; the same often applies to the addition of other nutrients. The maximum profit (V–C in Figure 8.3) is obtained from an amount of fertilizer, *F*, which gives a crop yield close to the maximum.

The actual amount of fertilizer to be added depends on the yield response per kilogram of applied nutrient, on the cost of fertilizer to the farmer and on the market price of the product. The yield response depends on the efficient use of fertilizer, discussed in later sections. Costs and prices vary between countries and are often influenced by government subsidies or price controls, which are used to increase food production.

For the subsistence farmer, who cannot afford financial risks, there are other considerations. He might have to borrow money at high interest rates to purchase fertilizer, the right fertilizer might not be available when needed, the market price of the product might be too low when he needs to sell, and the crop response might be low because of drought, excessive weed growth or insect attack.

Table 8.5. *Factors determining the growth*
and final yield of a crop

Climate:	Daily solar radiation
	Temperature
	Rainfall
	Daylength
Crop:	Crop variety
	Planting density
	Seedling emergence
	Control of pests, weeds, disease
Soil:	Available nutrients
	Presence of toxic substances, e.g. Al
	Soil physical conditions
	Soil depth
Fertilizer:	Application of required nutrients
	Application at correct time(s)
	Placement
	Application of correct amount

Fertilizers and crop response

The rate of growth and final yield of a crop depend on many factors of which nutrient supply and hence response to fertilizers is only one (Table 8.5).

The yields of wheat on Broadbalk field at Rothamsted Experimental Station, referred to above, illustrate an important principle. In the periods 1852–61 and 1970–8, the yield increased up to the highest rate of nitrogen application (144 kg N ha^{-1}) at which it averaged 5.32 t ha^{-1} between 1970 and 1978 but only 2.52 t ha^{-1} in the early period. Several changes were made during the conduct of the experiment. One was a change from autumn application of the fertilizer to spring application, but the biggest effect occurred when short-stemmed varieties were grown. Their harvest index (yield of grain/yield of grain plus straw) is higher than in the old varieties, that is, they transfer a higher proportion of photosynthate to the grain. Because of the strength of their stems they are less likely to lodge (fall over) when carrying heavy ears. The important principle is that at a particular site the response of crop yields to application of fertilizers depends on the standard of crop and soil management. Short-stemmed varieties of wheat and rice and appropriate use of fertilizer, irrigation and pesticides have been largely responsible for what has become known as the Green Revolution.

In areas of the world with low rainfall, water supply limits the crop

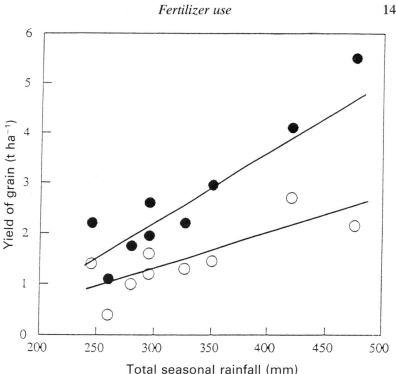

Figure 8.4 Yield of barley and its response to fertilizer increased with rainfall at experimental sites in Syria. Filled circles, with fertilizer, open circles, without fertilizer. (Derived from Shepherd, K.D. *et al.*, 1987. *Journal of Agricultural Science*, Cambridge **108**, 365–78.)

response to fertilizer application. Figure 8.4 shows that in Syria the response in grain yield of barley to addition of fertilizer increased with rainfall. With 300 mm of rain the response was 0.9 t ha^{-1}; with 450 mm, it was 2.1 t ha^{-1}. Because of the risk of crop failure in low-rainfall areas the use of fertilizer might not be advisable. Even where the rainfall is normally sufficient for high crop yields the response to fertilizer is reduced in a dry year. Nutrients in the fertilizer not taken up by the crop remain in the soil at harvest and those that remain soluble might subsequently be leached out of the soil.

Efficient use of fertilizers

The farmer aims for the maximum profit from his investment in fertilizers, which will usually mean that he requires a crop response close to the maximum.

Table 8.6. *The solutions commonly used to test for the adequacy of soil nutrient supplies*

Nutrient	Extracting solution
P	0.5 M sodium bicarbonate
P	0.03 M ammonium fluoride in 0.025 M hydrochloric acid
P	0.0125 M sulphuric acid in 0.05 M hydrochloric acid
K	1 M ammonium acetate[a]
K	1 M ammonium nitrate[a]
Micronutrient metals	0.05 M EDTA and other complexing agents

[a]Also used for Ca and Mg.

The principles of fertilizer use are now well developed and to obtain an optimum response:

1. It should be known which nutrients need to be added and the rate at which they should be added. Because the nutrients in fertilizers supplement those in soil, the soil supplies need to be known. Chemical analysis of the total content of the nutrients in soil provides almost no guidance. Instead, various acids, alkalis, salt solutions and complexing agents are used to extract soil samples, usually taken to a depth of 15–20 cm (Table 8.6). The amounts extracted are judged to be sufficient or deficient by comparing them with the amounts extracted from similar soils on which field experiments have tested the effects of increments of added nutrients on crop yield.

The field experiments themselves provide guidance to the farmer although the results strictly apply only to the site, crop, soil type and season in which the experiments were conducted. In addition to soil analysis, land use history can be used to predict the requirement for individual nutrients.

Crops may show shortages, excesses or imbalance of nutrients which, with experience, can be identified visually from symptoms which are characteristic of the disorder. For example, light green leaves will often indicate nitrogen deficiency, and chlorosis (yellowing) between the veins on older leaves will often indicate magnesium deficiency. Suspected deficiencies or excesses can be confirmed by chemical analysis of the plants or their parts, e.g. leaves, petioles or fruits at specific stages of growth. If it is too late to correct the disorder in the growing crop, it can be corrected for the next crop.

2. The fertilizer should be applied at the right time and at the right place. The critical periods when nutrients are most required are (i) soon

after germination, (ii) during the period of rapid vegetative growth, and (iii) for fruiting crops, at about the time of fruit production. Depending on the nutrient supply in the soil and the expected rainfall (which can cause leaching loss) the fertilizer might be given in one application or split into two or more applications.

Fertilizer can be broadcast on top of the soil, but it usually leads to better use by the crop if it is placed in the soil below the seed.

In the next section the properties of some of the individual fertilizers are described. The properties affect their utilization by crops and also have environmental consequences.

8.5 Individual fertilizers

When the required amount of each nutrient that needs to be added is known, the choice of fertilizer depends on the cost per unit mass of nutrient and on the nutrient being available for uptake by plants.

Nitrogen fertilizers

The requirement for fertilizer nitrogen depends on previous land use, the crop being grown, the standard of crop management and the expected yield. Use of fertilizer nitrogen is not justified if there is a sufficient supply of inorganic nitrogen (ammonium and nitrate) in the soil, as might occur after a period under grass or trees, or following the growth of a herbage legume. If available at low cost, organic manures such as farmyard manure are alternatives, but in practice the use of fertilizer nitrogen is necessary in many countries to provide the food crops required for animal and human consumption.

The main fertilizers that supply nitrogen are anhydrous ammonia (liquefied ammonia under pressure), ammonium nitrate, urea, and compound fertilizers supplying nitrogen and other nutrients. Their composition is given in Table 8.2.

World-wide, urea is the nitrogen fertilizer used in greatest amount. In soil, it rapidly hydrolyses, a reaction that is catalysed by the enzyme urease, which is released into soil by many microorganisms:

$$CO(NH_2)_2 + H_2O \xrightarrow{\text{urease}} CO_2 + 2NH_3,$$

and NH_4^+ and CO_3^{2-} are formed by further reaction with water.

Hydrolysis causes the pH to rise in the vicinity of the urea, and ammonia may be lost to the atmosphere by volatilization. To reduce or prevent this loss urea should be drilled into the soil 5–10 mm deep. Anhydrous ammonia should be injected into the soil for the same reason. Particularly large losses of ammonia can occur if urea is applied to the water in the cultivation of wetland rice. The water has a small buffer capacity compared with soils and its pH can exceed 10 for two reasons: (i) hydrolysis of the urea; and (ii) removal of carbon dioxide and bicarbonate ions by the algae during daytime which reduces the pH buffering.

A consideration with all fertilizers that supply ammonia, ammonium or urea is that nitrification causes the soil to become acid (Equations 9.31, 9.32 and 9.33). The cost of liming, where this is necessary to correct acidification, should therefore be added to the cost of the fertilizer.

Nitrate is readily leached from most soils, whether produced by nitrification of ammonium or added in fertilizers. It takes about 1–4 weeks for nitrification to be completed under good growing conditions, so that if heavy rain falls soon after application of the fertilizer, ammonium nitrogen, which is adsorbed by soil as an exchangeable cation, is conserved whereas nitrate nitrogen might be lost by leaching. Loss of nitrogen by denitrification (Section 5.7) occurs if nitrate ions are present under reducing conditions. This is the reason why fertilizer containing nitrate is not used for wetland rice.

Several proposals have been made to improve the efficiency of nitrogen fertilizers. They include (i) the inhibition of urease, which would prevent the hydrolysis of urea and allow it to be taken up directly by plants, (ii) the inhibition of nitrification, retaining nitrogen as ammonium which is normally not leached and which can be taken up by plants and (iii) the use of a range of slow acting fertilizers including sulphur-coated urea, isobutylidene diurea, and urea–formaldehyde products from the plastics industry. To date, none has justified the higher cost for agricultural crops. Where cost matters less, as for golf greens and ornamental plants, use is made of slow-acting fertilizers. In intensive horticulture some growers prefer to use organic fertilizers. They include dried blood and crushed hoof and horn, which contain nitrogen, and bone meal, which supplies nitrogen and phosphorus.

Phosphate fertilizers

Commonly, there is a deficiency of phosphorus when land is first brought into cultivation. As mentioned earlier, the most severe deficiencies have occurred on the old land masses of Africa, Australia and South America. Where there is a history of use of fertilizer phosphate the initial deficiencies will have been overcome; the requirement is then to maintain a sufficient supply of plant-available phosphate in the soil.

The choice of phosphate fertilizers is between those that are water-soluble, including single superphosphate, triple superphosphate (also known as concentrated superphosphate) and two ammonium phosphates (monoammonium and diammonium phosphate), and the water-insoluble rock phosphate (Table 8.2). Where transport costs are high, triple superphosphate is cheaper per unit of phosphorus than single superphosphate, but the latter contains sulphur and more calcium and may be preferred where deficiencies of these nutrients might also occur. The ammonium phosphates supply both nitrogen and phosphorus in concentrated form and are used in compound fertilizers.

Rock phosphate, which contains the mineral apatite, although not water-soluble, can be a useful fertilizer on acid soils for perennial crops that require only a low rate of supply of phosphate to the roots. The availability to plants of the phosphate varies between sources, rock phosphates of sedimentary origin being generally more available than those of igneous origin. Their value is usually assessed by their content of total phosphorus and the fraction that is soluble in dilute acids.

Single superphosphate is made by treating rock phosphate with sulphuric acid. Particularly in countries where sulphuric acid is expensive, an effective and cheaper fertilizer known as partially acidulated rock phosphate is made by using half the normal amount of sulphuric acid. Basic slag is another source of phosphate but is less important than in the past.

When water-soluble phosphate fertilizers are applied to soil, chemical reactions reduce their water solubility. Intermediate products include less soluble dicalcium phosphate and complex iron and aluminium phosphates. By further reaction under acid and neutral conditions phosphate ions become strongly held by hydrated oxides of iron and aluminium (Section 4.6), and above pH 7 calcium phosphates of low solubility, including apatite, are formed. Because of these reactions, phosphate fertilizers are usually more available for use by crop plants when they are placed below the seed.

The chemical reactions that decrease the concentration of phosphate in the soil solution also decrease its availability to plants. Experiments conducted in many countries have nevertheless shown that the water-soluble phosphate fertilizers have a residual availability that can last for several years. The effectiveness of the residues depends on the amount of phosphate initially applied, the amounts removed in the successive crops, and the buffer properties of the soil for phosphorus. In the United Kingdom the residual value of water-soluble phosphate fertilizers is assessed for the purpose of compensation to tenant farmers as half in the second year and one quarter in the third.

Potassium fertilizers

Deficiency of potassium can occur when new land is brought into cultivation but is more common after a period of cultivation. Leafy crops and fruit crops can remove over 100 kg ha^{-1} of potassium when they are harvested, thus depleting the reserve of available potassium in the soil. If present in the soil, the clay mineral illite will slowly release potassium, as will the primary minerals orthoclase feldspar and micas, though more slowly. However, the release is usually not fast enough to meet the requirement of rapidly growing crops.

The only two potassium fertilizers in common use are the chloride and the sulphate. The chloride is cheaper per unit of potassium and is used in much greater amounts. The sulphate is often preferred by horticulturists in the belief that it gives better-quality crops, for which there is some evidence for the potato crop. The chloride is not used in growing tobacco because it gives the cured crop poor burning quality.

The correction of other nutritional problems

In the United Kingdom, agricultural soils that have no reserves of calcium carbonate are regularly limed to keep the pH (in water) close to 6.5. If the pH is allowed to fall to 5.5 or below, many crops fail due to toxicity of aluminium and some to manganese. In very strongly leached and acid soils crop failure may be due to calcium deficiency, which can be corrected by adding a calcium salt such as gypsum.

Additions of magnesium sulphate, or calcium or magnesium carbonate if liming is needed, are most likely to give crop responses on leached, sandy soils. Fruit trees and sugar beet may show signs of

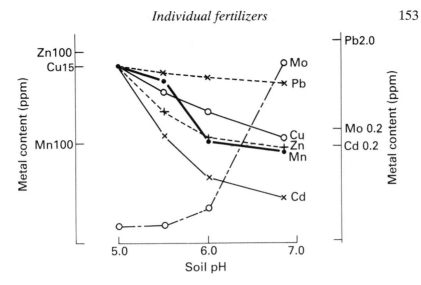

Figure 8.5 Relation between soil pH and the concentration of six metals in subterranean clover: pH did not affect yield. (From Williams, C.H., 1977, *Journal of the Australian Institute of Agricultural Science* **43,** 99; with permission, AIAS copyright.)

magnesium deficiency, as do tomato crops if excessive potassium manuring has resulted in a high K : Mg ratio in the soil.

Sulphur deficiency has increased in industrialized countries during the past ten years because less sulphur dioxide is being emitted into the atmosphere with the result that less is deposited onto soil and crops. Sulphur can be added as calcium sulphate (gypsum), in ammonium sulphate, or as single superphosphate when phosphate is also needed.

Deficiencies of the micronutrients are most common on old land surfaces and tend to occur with sensitive crops and particular soil conditions. For example, the solubility of most micronutrients decreases as the pH is raised, although the solubility of molybdenum increases. The uptake by plants follows the same pattern (Figure 8.5). Deficient supplies of iron, manganese and boron are most likely to occur in soils of high pH. Maize and rice appear to be particularly sensitive to zinc deficiency, oats and some varieties of wheat to copper deficiency, and sugar beet, swedes and cotton to boron deficiency. All organisms that fix nitrogen require molybdenum; in Australia, New Zealand and elsewhere the successful establishment of clovers has depended on the addition of molybdenum.

Table 8.7. *Effects of organic matter on soil fertility*

Physical:	increases supply of water to crops
	increases aggregation of soil particles
	may improve drainage and hence early growth of crop
	gives greater flexibility for cultivations
Chemical:	releases N, P and S on mineralization
	retains nutrient cations, e.g. Ca^{2+}, Mg^{2+}, K^+, NH_4^+, against leaching loss
	chelates micronutrients, generally increasing uptake by plants
	acts as a pH buffer
	reduces the hazard from heavy metals
Biological:	may support organisms which help to control root diseases

8.6 Organic matter and organic manures

The sources of soil organic matter under natural conditions are plant residues and litter from trees. In soil which is cultivated organic matter accumulates under bush or grass fallows and is added in crop residues, weeds and organic manures.

There are several ways in which organic matter might benefit an arable crop (Table 8.7). Field experiments in the UK have generally shown that the main benefit is from the provision of nitrogen when the organic matter mineralizes after a grass fallow, applications of farmyard manure or applications of peat.

The results of field experiments at a site in southern England (Figure 8.6) were obtained on paired plots, one member of each pair having received applications of peat to raise the content of soil organic matter. Yield of winter wheat was only higher after application of peat on plots receiving no fertilizer nitrogen whereas the yield of potatoes was higher at all four rates of application of nitrogen. The results with winter wheat are consistent with the explanation that the response was due to the mineralization of nitrogen from peat. With potatoes there was an additional effect, which might have been physical. Organic matter increases the ability of soil to retain water available to plants, and although the difference in organic matter content of the plots was small, it might have been sufficient to provide two or three days' supply of water. This would explain why the effect of peat was greater with potatoes, a shallow-rooted crop, than with winter wheat, which roots more deeply.

In addition to supplying nutrients the other effects of organic matter (Table 8.7) might be important in different circumstances to those of the experiment described above. In particular, organic matter helps to

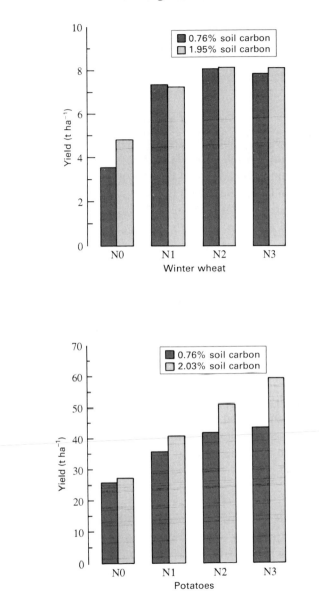

Figure 8.6 Yields of winter wheat and potatoes on paired plots at two levels of organic matter (by addition of peat) and receiving four rates of fertilizer nitrogen (N0, N1, N2, and N3). (From Johnson, A.E., 1990. *Fertilizer Society of South Africa Journal*, **1,** 10–19.)

stabilize aggregates and keep open the macropores needed for infiltration, which will reduce run-off and erosion. Although the reasons are not always known, it is commonly found that soils are easier to manage, especially in the tropics, when the organic matter content is high.

Farmyard manure

This consists of faeces, urine and bedding material, which is usually cereal straw. Its composition depends on the domestic animals (usually cattle or horses) being kept, the feed they receive and bedding materials used. Nutrients can be leached from it if it is exposed to heavy rain and ammonia can be lost by volatilization. Its composition is therefore variable; some typical analyses of cattle and other organic manures are given in Table 8.8. Additionally it contains calcium, magnesium, sulphur and micronutrients. It supplies all nutrients, although additional nitrogen is usually needed to balance the nutrient requirements of crops. In the year of application about one third of the nitrogen, one half of the phosphorus and all the potassium become available for crop uptake. In an application of 25 t ha^{-1} of cattle manure and a composition of 0.5% N, 0.15% P and 0.6% K (Table 8.8) the amounts *available* in the year of application are about 42, 19 and 150 kg per hectare of N, P and K respectively.

Farmyard manure also improves soil structure; if applied frequently it increases the content of soil organic matter (Figure 3.1*b*). Because it decomposes slowly in soil the nutrients not available to crops soon after application become available in later seasons.

Animal slurries

These are the semi-liquid form of faeces, urine and floor washings from housed cattle and pigs. It was estimated that in the early 1980s in England and Wales about 70% of the droppings of cattle and 50% of the

Table 8.8. *Some typical analyses of organic manures (on fresh mass)*

	% dry matter	%N	%P	%K
Poultry manure	70	2–4	1–1.5	1–1.5
Cattle manure	25	0.5–0.6	0.15	0.6
Cattle slurry	10	0.5	0.1	0.4
Pig slurry	2–10	0.3–0.6	0.1	0.2

Table 8.9. *Distribution of total nitrogen^a between sinks after application of cattle slurry to grassland in winter and spring*

Figures given are distributions of N as a percentage of application.

Application of slurry	Apparent recovery in herbage^b	NH$_3$ volatilization	Denitrification loss	Total of sinks
Winter experiment				
Surface spread	20	31	12	63
Injected	33	1	21	55
Injected + nitropyrin	36	1	9	46
Spring experiment				
Surface spread	26	20	2	48
Injected	36	1	7	44
Injected + nitropyrin	42	1	5	48

^{a}Application rate of total N was 248 kg ha^{-1} in the winter experiment and 262 kg ha^{-1} in the spring experiment.
^{b}Calculated from the nitrogen taken up by the herbage compared with that in plots not receiving slurry.
Source: From Thompson, R.B., Ryden, J.C. & Lockyer, D.R. 1987. *Journal of Soil Science* **38**, 689–700; with permission.

droppings of pigs went into slurry; the rest was incorporated with the bedding material as farmyard manure.

The liquid slurry is collected in tanks for distribution between late autumn and early spring. It is usually sprayed onto grassland, or onto bare soil before planting a spring-sown crop. Unless a surface application is followed by heavy rainfall, about half the NH_4^+-N that it contains can be lost by volatilization. If it is injected into the soil, the combination of a high volume of liquid and the presence in it of energy-rich organic substances leads to reducing conditions; nitrate present in the soil may then be denitrified. These points are illustrated by the results of two experiments using slurry from dairy cattle on grassland, given in Table 8.9. In these experiments 42% of the nitrogen applied was present as NH_4^+-N and the rest was in organic compounds. The experiments included a treatment with nitropyrin, which inhibits nitrification, and it will be seen to have reduced the loss of nitrogen by denitrification when the slurry was injected. There was negligible leaching loss and the nitrogen not accounted for was assumed to have been added to the stock of organic nitrogen in the soil.

Cattle and pig slurries supply all the essential plant nutrients. When they are being sprayed on fields they do, however, have an offensive smell.

The use of sewage sludge is discussed in Section 10.3.

8.7 Nutrient balances

The term 'nutrient balance' is used for the comparison between total input of each nutrient and total output. If the output exceeds the input, soil nutrient supplies will become depleted.

In the nutrient balance shown in Table 8.10, the removal of nitrogen and potassium in four crops exceeded the additions in farmyard manure and fertilizers, and the removal of phosphorus was less than the additions. A complete balance sheet would, however, need entries for the addition of nutrients from the atmosphere in rain, dry deposition, by biological nitrogen fixation and by mineral weathering, and for losses in drainage water and by volatilization.

Loss of nutrients in drainage water is shown in Table 8.11, which gives concentrations in drainage water from arable land at three experimental sites. Of the major nutrients, it will be seen that losses of phosphate and potassium were small and that nitrogen was present in the drainage water almost entirely as nitrate. Losses of calcium, magnesium and sulphate were large.

Establishing a nitrogen balance has proved to be particularly difficult because (i) there are large amounts of organically held nitrogen in soil and small changes in the amount cannot easily be measured, and (ii) it is difficult to measure the dry deposition of nitrogen compounds from the

Table 8.10. *The nutrient balance in a four-course rotation (kale, barley, ryegrass, wheat) in an experiment at Rothamsted Experimental Station*

| | Nutrient (kg ha^{-1}) | | |
	N	P	K
Added in farmyard manure	56	34	168
Added in fertilizers	359	81	151
Total additions	415	115	319
Removed in four crops	437	58	348
Balance	−22	+57	−29

Source: From Cooke, G.W. 1967. *The Control of Soil Fertility.* Crosby Lockwood, London.

Table 8.11. *Average concentrations of elements (mg l⁻¹)ᵃ in water from land drains at three sites in southern and eastern England*

Ca	168	NO_3-N	17
Mg	7.2	SO_4-S	51
K	1.8	Cl	47
Na	16	PO_4	0.06
NH_4-N	0.5		

[a]The numbers also represent amounts in kg ha⁻¹ if the amount of drainage water is 100 mm.
Source: From Cooke, G.W. 1976, *Agriculture and Water Quality*. MAFF Technical Bulletin **32**. HMSO, London.

Table 8.12. *Nitrogen balance in soil growing winter wheat, Broadbalk, in 1980 and 1981*

	kg N ha⁻¹
Inputs	
Fertilizer	189
From external, non-fertilizer sources	50
From soil organic matter	59
Total	298
Outputs	
To grain	136
To straw	33
To soil organic matter	59
Losses to drains and atmosphere	70
Total	298

Source: From Jenkinson, D.S. and Parry, L.C. 1989.
Soil Biology and Biochemistry, **21**, 535.

atmosphere to soils and crops, and the flow of nitrogen compounds to the atmosphere. An estimate for the Broadbalk Field at Rothamsted Experimental Station is given in Table 8.12.

Nutrient balances are rarely known with any degree of accuracy because of lack of information on leaching losses, inputs from the atmosphere and by mineral weathering, and outputs to the atmosphere. They are useful in showing if removal in crops or animals exceeds the additions in fertilizers and manures, but soil analysis every two or three years, and checking crop yields and signs of deficiencies, are a better guide that nutrient supplies are being maintained.

8.8 Water

Arable and grass crops require water throughout their period of growth. If the supply is inadequate, transpiration and then growth rate decrease, and the final yield is reduced.

The amount of water transpired by plants growing with a non-limiting supply of water and forming a complete canopy is known as the potential transpiration. It can be calculated from meteorological measurements of temperature, relative humidity, wind velocity and solar radiation (Section 6.6). Actual transpiration is less than the potential when the plant canopy is incomplete, for example during the early growth of an annual crop, and also when the plants are stressed by water shortage.

A comparison between potential transpiration and rainfall at a site in West Africa is shown in Figure 8.7. The average annual rainfall is 882 mm, which falls over 7 months with almost none in the rest of the year. A crop is sown when there has been about 25 mm rain in a ten-day period, which occurs on average at this site in the middle of May. Rainfall exceeds potential transpiration from late June, when the canopy will be complete, until the middle of September. During this period water is stored in the soil, up to a capacity of 100 mm at the site, and excess drains through the soil. After the middle of September

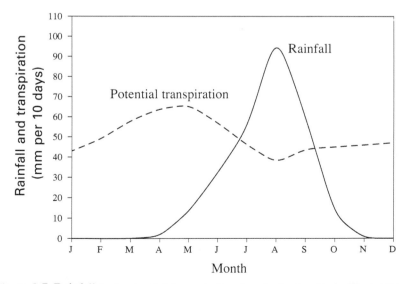

Figure 8.7 Rainfall and potential transpiration by plants at a site in West Africa with a dry season of five months.

rainfall continues to decrease and the crop becomes increasingly dependent on stored water, which is used up by the end of October.

The average rainfall and length of the rainy period make the site suitable for growing millet, sorghum, groundnuts and short-season maize. South of the site the rainfall is higher; to the north it is less, and less reliable, the effects of which are discussed in Section 13.6.

8.9 Crop production in low-rainfall regions

Several land management systems have been developed to provide food in regions with low rainfall. Their purpose is either to use rainfall as efficiently as possible or to increase the water supplies by irrigation. Rain-fed farming will be considered first.

Where rainfall limits crop production the farmer might (i) conserve water by minimizing loss by run-off and evaporation from the soil, or (ii) increase the total supply to the crop (Table 8.13), and he might do both.

Run-off can be minimized by using terraces. Usually the retaining banks are graded to run slightly off the contour so that the rate of run-off is reduced, thus allowing more water to infiltrate the soil (Section 12.8). In a ridge-and-furrow system the furrows are aligned similarly. Absorption terraces with the retaining banks on the contour are suitable only if there is no danger of heavy storms causing flooding and erosion. Infiltration should be kept as high as possible by cultivating the soil surface or, if the material is available, by applying a mulch of plant residues.

Another benefit of a mulch is that it reduces the loss of water by evaporation from a bare, wet soil. A plant canopy also protects wet soil

Table 8.13. *Methods used to optimize the supply of water to crops in dryland farming*

1. *Reduce run-off*	contour cultivation
	use absorption terracing
	use crop residues as mulch
	timely cultivations to increase infiltration
2. *Reduce evaporation from soil*	establish crops early
	use mulches
	supply fertilizer where rainfall is reliable
3. *Increase water supply to the crop*	control weeds
	collect water during a bare fallow
	rain harvesting from a catchment
	grow deeper-rooted crops

from evaporative loss, so that early establishment and the use of fertilizers to obtain rapid growth can enhance the efficiency of water use.

Bare fallowing is a practice used in low rainfall areas to increase water storage. If the land is kept free of weeds during a summer fallow, water from rainfall in the previous winter is not lost by transpiration although some is lost by evaporation, as is some of the rain that falls during the summer. The fraction of rain that is stored by a summer fallow has been found to depend on the effective control of weeds and run-off. It also depends on there being a sufficient amount of rain to penetrate the soil because light showers are lost by evaporation. Retaining the stubble (stubble mulching) increases water storage in the Great Plains of the USA, mainly because it holds the winter snow.

Rain harvesting has long been used to divert run-off from stony hill slopes into valley sites with permeable soils. In the deserts of the eastern Mediterranean region a ratio of collecting area to cultivated area of about 25 : 1 was used in the past for successful crop production. An experimental farm in Israel is using the same system and variants are being tried out in some African countries.

8.10 Irrigation

Application of water to the land is one of the oldest techniques to be used to ensure adequate food supplies. In the shadoof system in the Nile Valley, a bucket held on a pole is filled with water from a river and is raised by lowering a counter-weight. The water is tipped into a gully and led into the field to be irrigated. A faster method is to raise the buckets on a water wheel which is rotated by an ox or water buffalo. Another traditional method is to allow a river to overflow and flood portions of land surrounded by banks (bunds) of soil.

Most irrigation schemes in recent years have depended on the construction of dams behind which there are raised water levels. When required the water is fed into fields via a system of canals and channels. Where there are no local rivers or lakes the water table can be tapped by sinking a borehole.

There have also been changes in the way the irrigation water is applied. Surface application into basins or furrows using gravity depends on the land being almost flat, and on the soil having a low infiltration rate in order to ensure reasonably even distribution. Overhead sprinkling overcomes these problems but requires the water to be pumped. Trickle irrigation, a system introduced in recent years, gives the greatest water use efficiency (units of harvested crop per unit of water used). The water is fed into plastic pipes with a narrow hole near each plant or tree through which it slowly drips.

The operation of an irrigation scheme requires skill. Water is often a scarce and expensive commodity and must be used efficiently, which means that it must be used in the right amounts and at the right times. The farmer might make decisions on when to apply water based on experience or signs of wilting of the crop. Alternatively, water deficits in the soil can be estimated from open-pan evaporimeters or from the calculated evapotranspiration loss (see also Section 6.6). Water is then applied when the deficit reaches a value found to be critical in field experiments. Usually, irrigation water is applied to supplement the rainfall and if it rains after the water has been applied the soil can become waterlogged.

The area of land under irrigation has increased substantially in recent years. Between 1950 and 1985 the area increased almost three-fold (Figure 8.8), and about one sixth of the arable land of the world is now

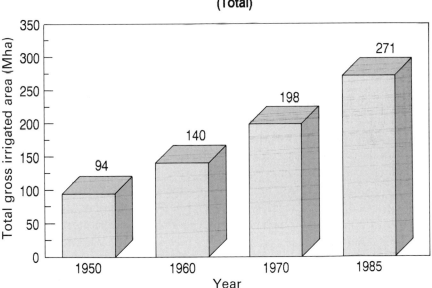

Figure 8.8 Increase in irrigated area through the world, 1950–85. (From Rangeley, W.R. 1986. *Philosophical Transactions of the Royal Society of London* A**316,** 347–68.)

Table 8.14. *Salinity and sodicity in irrigated soils*

1. Addition of water	Assuming 50 cm is needed during the growing season, total addition per hectare is 50×10^5 l (5000 m^3).
2. Addition of salt	(*a*) If the water contains 500 mg l^{-1} of dissolved salt, the addition in 50 cm is 2500 kg ha^{-1}. The salt concentration is usually specified by the electrical conductivity of the water.
	(*b*) The sodium hazard from the dissolved salts is expressed as the sodium adsorption ratio (SAR) where SAR = [Na]/√[Ca + Mg] with concentrations in millimoles per litre. If the water contains 150 mg l^{-1} of Na (6.52 mmol l^{-1}), 5 mg l^{-1} of Ca (0.125 mmol l^{-1}) and 5 mg l^{-1} of Mg (0.208 mmol l^{-1}) the SAR is 11, which is a medium sodium hazard.
3. Soil salinity	(*a*) Total salinity is usually specified by the electrical conductivity of the saturation extract of the soil. Crops that are sensitive to salt are affected if the conductivity exceeds 4 deci siemens per metre (4 dS m^{-1}).
	(*b*) Exchangeable sodium becomes a hazard if it exceeds about 15% of the total exchangeable cations.

Note: Because of the hazards, salinity and sodicity have been the subjects of intensive research. Books suggested for further reading should be consulted for the use of the measurements listed above. It should also be noted that salt concentrations are used above whereas activities are more properly required.

irrigated. It has been estimated that this one sixth of the land area produces one third of the world food production. About half the land used for growing rice is irrigated.

A major problem with irrigation in regions with low rainfall is the accumulation in soil of salts, especially sodium salts, which can then render the soil too saline for crop growth. The sources of the salts are the irrigation water itself, and saline groundwater which rises toward the soil surface when excess water is applied and there is no drainage system. Irrigation channels that leak can also cause serious soil salinity.

Soil salinity is usually confirmed by measuring the electrical conductivity of a saturated extract of soil; the limits for crop growth are given in Table 8.14.

To prevent the accumulation of harmful concentrations of salts, and to remove the salts if they have accumulated, it is necessary to add excess water. The salts are washed into the subsoil and conducted via drains into a river. The requirement for a drainage system adds substantially to the cost of an irrigation scheme.

Sodium salts can also increase the exchangeable sodium content of the soil, giving rise to sodicity. If the content exceeds about 15% of the exchangeable cations the clay disperses when the salts are washed out. The soil pores become blocked with the dispersed clay and the permeability to water is reduced. The problems of salinity and sodicity are particularly acute in semi-arid regions.

8.11 Summary

As in natural ecosystems (Chapter 7), physical and chemical conditions in the soil affect the development of the root systems of crops. The difference between the two is that the conditions can be improved for the production of crops. This is done by drainage to remove surplus water, cultivations to improve the soil mechanical properties, application of nutrients in fertilizers and organic manures, use of lime to correct acidity, and by irrigation.

Removal from the land of crops (arable crops and trees) and animal products depletes the stock of nutrients in the soil, and cultivations may lead to deterioration of soil structure. To maintain fertility (the capacity of the soil to support the crop being grown) nutrients need to be added, addition of organic matter usually improves soil structure and helps to increase crop yields, and lime might be needed to maintain the pH.

Successful long-term management of soils is needed if the food requirements of a rapidly increasing world population are to be met.

9

Soil acidification

9.1 Introduction

Most soils will become acidic if they are exposed for a sufficient length of time to rainwater which causes through-drainage. Acidification is a natural process and is referred to in Section 3.5 as one of the processes of soil formation. Evidence of soil acidification is found in rocks formed 300 million years ago in the Carboniferous period, which contain fossil soils similar to spodosols of the present day, and these are acidic. Roman writers of more than 2000 years ago described the benefits of liming, suggesting that acid soils were then a problem. At the present day all the continents of the world contain large areas of acid soils.

In recent years, however, pollutants in the atmosphere have increased the rate of acidification of soils and fresh waters, and there has been concern that this has caused the death of trees and fish. The acidifying effect of atmospheric pollutants is often described under the title 'acid rain' although, as will be described in Section 9.7, this can be misleading.

This chapter deals with the process of soil acidification and the environmental effects of acid soils. A little chemistry is needed if the processes are to be understood, and we start by explaining what is meant by acidity, and soil acidity in particular.

9.2 pH and buffering

To avoid confusion, definitions are needed of the terms to be used. As we are dealing with aqueous solutions, pure water provides the starting point.

Water molecules are dissociated to a very small extent:

$$H_2O \rightleftharpoons H^+ + OH^-, \tag{9.1}$$

giving hydrogen ions (protons) and hydroxyl ions in equal concentrations of 10^{-7} M at 25 °C. Instead of using concentrations, which are cumbersome because of their wide range in acid and alkaline solutions, it is convenient to refer to the negative logarithm of the concentration, known as the pH:

$$pH = -\log[H^+]. \tag{9.2}$$

Strictly, pH is the negative logarithm of the hydrogen ion activity rather than the concentration, but in dilute solutions the two can be assumed to be equal. A further point is that H^+ reacts with H_2O to form H_3O^+ (the hydronium ion). This does not affect the definition of pH, and the formation of H_3O^+ will not be considered further in this text.

Acid and base

An acid donates protons (hydrogen ions) and a base accepts protons:

$$acid \rightleftharpoons base + proton, \tag{9.3}$$

for example

$$HCl \rightleftharpoons Cl^- + H^+; \tag{9.4}$$

$$NH_4^+ \rightleftharpoons NH_3 + H^+. \tag{9.5}$$

The neutral point is pH 7.0. A solution is acidic if the pH is below 7 and alkaline if it is above 7. A solution of pH 5 has a concentration of 10^{-5} M hydrogen ions, and one of pH 8 has a concentration of 10^{-8} M hydrogen ions and 10^{-6} M hydroxyl ions.

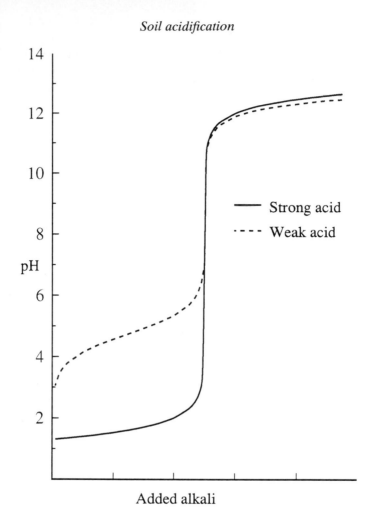

Figure 9.1 Titration curves of a strong acid (hydrochloric acid) and a weak acid (acetic acid) with sodium hydroxide solution.

Buffer capacity

This is the amount (mol l^{-1}) of acid (H^+) or base (OH^-) required to change the pH of 1 litre of solution by 1 unit. If a dilute solution of hydrochloric acid is titrated with sodium hydroxide the pH rises sharply above 3 (Figure 9.1); this is because it exists almost completely as H^+ and Cl^-, at least in dilute solution. Similar titration of acetic acid gives only a small change of pH when alkali is added at about pH 5. The acetic acid buffers the pH at about 5 because its protons are tightly bound and are only released when sufficient alkali is added:

$$CH_3COOH \rightleftharpoons CH_3COO^- + H^+. \qquad (9.6)$$

Generally for a weak acid

$$HA \rightleftharpoons H^+ + A^-. \qquad (9.7)$$

Its degree of dissociation into ions is expressed as

$$K_A = \frac{[H^+][A^-]}{[HA]}, \qquad (9.8)$$

where K_A is known as the dissociation constant of the weak acid. It is usually expressed as the negative logarithm (pK_A), by analogy with pH. The units of concentration are moles per litre.

If half of HA is dissociated, $[HA] = [A]$, so that $K_A = [H^+]$, or $pK = pH$; that is, the pK is the pH at which half the acid is dissociated. This is also the pH at which the acid is most strongly buffered. The pK of acetic acid is 4.76, and Figure 9.1 shows that buffering is strongest at this value of pH.

The two examples shown in Figure 9.1 are of a strong acid and a weak acid. There is a third type of acid known as a Lewis acid; Al^{3+} and Fe^{3+} are examples of Lewis acids. In aqueous solution these ions are hydrated as $Al(H_2O)_6^{3+}$ and $Fe(H_2O)_6^{3+}$ respectively. Protons can be dissociated from the water molecules:

$$Al(H_2O)_6^{3+} \rightleftharpoons AlOH(H_2O)_5^{2+} + H^+ \qquad (9.9)$$

and

$$K_1 = \frac{[Al(OH)^{2+}][H^+]}{[Al^{3+}]}, \qquad (9.10)$$

where $[Al^{3+}]$ represents $[Al(H_2O)_6^{3+}]$ and $Al(OH)^{2+}$ represents $AlOH(H_2O)_5^{2+}$. Further protons can be sequentially removed with the formation of $Al(OH)_3$. The intermediates include polymers of uncertain composition, but ignoring these

$$AlOH(H_2O)_5^{2+} \rightleftharpoons Al(OH)_2(H_2O)_4^+ + H^+; \qquad (9.11)$$

$$Al(OH)_2(H_2O)_4^+ \rightleftharpoons Al(OH)_3 + H^+ + 3H_2O. \qquad (9.12)$$

In logarithmic form, pK_1 (Equation 9.10) $= 5$, and following the reasoning above, at pH 5 the concentrations of $Al(OH)^{2+}$ and Al^{3+} are equal.

Protons are released if base is added to Al ions and by the reverse reaction protons are removed from solution when acid is added to $Al(OH)_3$. This is one of the important processes that buffers the change of pH in soils. Ferric ions enter pH buffer reactions at lower pH than aluminium, the value of pK_1 being 2.2.

Because of the buffering properties of many solutions, pH alone is not sufficient to define their acidity. The term 'base neutralizing capacity' (BNC) is used to describe the amount of base that needs to be added to reach some reference point, say pH 7. Similarly 'acid neutralizing capacity' (ANC) describes the amount of acid needed to reach the reference point.

9.3 Soil pH and buffer capacity

The principles described above also apply to soil, but there are other considerations because of the ion exchange properties of clay and humus, and for other reasons.

Soil pH can be measured on the extracted soil solution, or on a suspension of soil in water. For reasons given in Section 4.4 the pH should strictly be measured when the soil is suspended in a solution of about the same composition as the soil solution. Commonly, however, the pH is measured on a suspension in water.

The buffering of soil pH is due to several soil properties, the effect of which is to make the drop in pH less than in an equivalent volume of water when acid is added. These properties are listed in Table 9.1. The reactions that occur with added acid are summarized as follows:

1. Carbonates are dissolved by carbonic acid and other added acids:

$$CO_2 + H_2O \rightleftharpoons H_2CO_3 \rightleftharpoons H^+ + HCO_3^-; \qquad (9.13)$$

$$CaCO_3 + H^+ + HCO_3^- \rightleftharpoons Ca(HCO_3)_2. \qquad (9.14)$$

The product, calcium bicarbonate, is soluble and is leached out of the soil by excess rain.

Table 9.1. *The pH buffer properties of soils*

---
1. Reaction of acids with calcium and magnesium carbonates
2. Cation exchange
3. Proton adsorption by clay minerals, humus, and hydrated aluminium and iron oxides
4. Proton adsorption by aluminium ions
5. Solubilization of soil minerals
---

2. Cation exchange, which removes protons from solution:

$$M^+ \text{ soil} + H^+ \rightleftharpoons H^+ \text{ soil} + M^+, \qquad (9.15)$$

where M^+ is an exchangeable cation. Exchangeable H^+ reacts with aluminium in clays and hydrous oxides (reactions 4 and 5 below).

3. Humic acid:

$$-COO^- + H^+ \rightleftharpoons -COOH, \qquad (9.16)$$

and at a high pH,

$$> CO^- + H^+ \rightleftharpoons > COH. \qquad (9.17)$$

The effect of these reactions is to reduce the negative charge on ionized carboxyl and phenolic hydroxyl groups, respectively (see Section 2.8).

4. Aluminium buffering

$$AlOH(H_2O)_5^{2+} + H^+ \rightleftharpoons Al(H_2O)_6^{3+}. \qquad (9.18)$$

See equations 9.9, 9.11 and 9.12.

5. Proton adsorption by clays and oxides of Al and Fe:

$$> AlO^- + H^+ \rightleftharpoons > AlOH; \qquad (9.19)$$

$$> AlOH + H^+ \rightleftharpoons > AlOH_2^+. \qquad (9.20)$$

These reactions lower the pH-dependent negative charge and are the source of positive charge (Section 4.1).

6. Hydrolysis of minerals:

$$4KAlSi_3O_8 + 4H^+ + 18H_2O \rightarrow Si_4Al_4O_{10}(OH)_8 + 4K^+ + 8Si(OH)_4. \quad (9.21)$$

In this reaction orthoclase feldspar forms kaolinite and silicic acid (Section 3.2). Hydrolysis of other silicates also removes H^+ from solution.

In all these reactions soil components remove protons from the acidified solution and the pH will decrease to an extent that depends on the buffer capacity of the soil. The reactions are discussed further in Section 9.5 in relation to the processes of acidification.

9.4 Percentage base saturation

The cations Ca^{2+}, Mg^{2+}, Na^+ and K^+ are known as basic, or base, cations. The negative charge on clay and humus is balanced by the charge on these cations and also by H^+ and aluminium ions (Al^{3+}, $AlOH^{2+}$, $Al(OH)_2^+$) which are known as acidic cations.

$$\text{Percentage base saturation} = \frac{\Sigma(\text{exchangeable basic cations})_c}{(\text{cation exchange capacity})_c} \times 100. \qquad (9.22)$$

The cation exchange capacity is measured at pH 7, sometimes at a higher pH. The use of CEC at pH 7 is unrealistic for acid soils with dominantly pH dependent charge. For these soils, which are common in tropical regions, a more useful definition is:

$$\text{Percentage base saturation} = \frac{\Sigma(\text{exchangeable basic cations})_c}{\Sigma(\text{total exchangeable cations})_c} \times 100. \qquad (9.23)$$

A solution of unbuffered KCl is used to displace the exchangeable cations at the pH of the soil. The displaced H^+, Al^{3+}, $Al(OH)_2^+$ are collectively known as exchangeable acidity. The total exchangeable cations measured by this method is the effective cation exchange capacity of the soil, ECEC, and is referred to in Section 4.2. The percentage base saturation is used, in addition to pH, to indicate whether soil acidity will affect plant growth.

9.5 Processes of soil acidification

In a neutral soil the exchangeable cations that dominate the exchange capacity are Ca^{2+}, Mg^{2+}, K^+ and Na^+ (the basic cations). As a result of acidification these basic cations become replaced by protons and aluminium ions. The loss of basic cations is permanent if they are leached out of the root zone or are removed in a harvested crop. The loss is temporary if they are taken up by plants and returned to the soil in litter or on the death of the plants. Some of the basic cations taken up by trees are held in woody tissues where they are effectively removed from the soil for many years. A distinction therefore needs to be made between acidification of soil and acidification of the soil–plant ecosystem. Sources of acidity (and alkalinity) are listed in Table 9.2, and Figure 9.2 gives a schematic representation of the processes.

1. Effects of the atmosphere

Deposition from the atmosphere can cause soil acidification in several ways:

(i)   Acids and acid-forming chemicals in precipitation (rainfall, snow and fog) falling on to plants and soil.

Table 9.2. *Sources of acidity and alkalinity in soils*

Acidity
1. Addition of acids and acid-forming chemicals from the atmosphere
2. Uptake of $(cations)_c$ > $(anions)_c$ by plants[a]
3. Removal of basic cations in plant harvest and by leaching
4. Oxidation processes such as nitrification
5. Microbial production of organic acids
6. Increase of soil organic matter content
7. Volatilization of ammonia from ammonium compounds

Alkalinity
1. Addition from the atmosphere of carbonates and weatherable minerals in dust
2. Uptake of $(anions)_c$ > $(cations)_c$ by plants
3. Reduction processes
4. Weathering of primary minerals
5. Hydrolysis of exchangeable sodium ions, which may be present in high concentrations after irrigation with saline water

[a]$(cations)_c$ is the electrical charge on total cations, and $(anions)_c$ is the electrical charge on total anions.

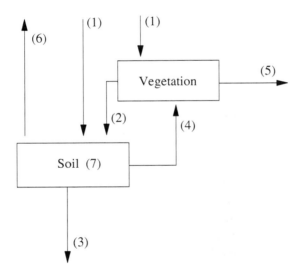

Figure 9.2 Processes that affect the acidification of ecosystems. 1, wet and dry deposition; 2, throughfall and litterfall; 3, leaching; 4, nutrient uptake; 5, nutrient removal in harvest; 6, volatilization of ammonia; 7, soil processes (see text).

(ii)   Absorption of gases from the atmosphere by plants and soil (dry deposition).
(iii) Deposition of aerosols (occult deposition) onto plants and soil.

(iv)   Organic acids washed by rain from the leaf canopy (throughfall) and
       down the stems of plants (stem flow).

   The individual components that cause acidification fall into three
groups:
   (i)   acids in rainwater: H_2CO_3, HNO_3, H_2SO_4, HCl.
   (ii)   acid gases: CO_2, SO_2, SO_3, HNO_3. In soil water:

$$CO_2 + H_2O \rightleftharpoons H_2CO_3; \qquad (9.24)$$

$$SO_2 + H_2O \rightleftharpoons H_2SO_3 \rightarrow H_2SO_4; \qquad (9.25)$$

$$SO_3 + H_2O \rightarrow H_2SO_4. \qquad (9.26)$$

(iii)   acid-forming chemicals: NH_3 and $NH_4{}^+$. In soil

$$NH_3 + 3O \xrightarrow[\text{oxidation}]{\text{microbial}} NO_2^- + H^+ + H_2O; \qquad (9.27)$$

$$NH_4^+ + 3O \xrightarrow[\text{oxidation}]{\text{microbial}} NO_2^- + 2H^+ + H_2O; \qquad (9.28)$$

followed by $$HNO_2 + O \xrightarrow[\text{oxidation}]{\text{microbial}} HNO_3. \qquad (9.29)$$

The reactions represented by Equations 9.27–9.29 describe nitrification,
which causes acidification of soils; they are referred to again below.
   Hydrochloric, nitric and sulphuric acids are strong acids and can
lower the pH of rainwater below 5.6, the value of pure water in
equilibrium with an atmosphere containing 340 ppmv of CO_2. The
sources of the strong acids are referred to in Section 9.7.

2. Effects of plants

If plants take up more cations (strictly, cation charge) than anions
(anion charge) they synthesize more organic acids and excrete protons
from their roots (Section 7.6); this process lowers the pH. Uptake of
excess anions raises the soil pH. The elements required in largest
amounts from soil are nitrogen and potassium. Nitrogen can be taken up
by plants as NH_4^+ or NO_3^- and potassium is taken up as K^+. If NH_4^+
rather than NO_3^- is taken up by plants the uptake ratio of cation
charge: anion charge will usually be greater than 1 and acidification will

occur in the root zone. Acidification is also caused by legumes, which obtain their nitrogen by biological fixation and take up excess cations.

The soil acidification caused by an uptake ratio greater than 1 will be temporary if all the plant material is returned to the soil. This is because the organic acids decompose to carbon dioxide and water, and the basic cations are released into the soil. Release of the basic cations counterbalances the acidification that occurs during plant uptake. If plants or trees are harvested acidification results from the removal of the basic cations from the site, whether or not cation uptake exceeded anion uptake.

3. Oxidation–reduction processes

Oxidation and reduction processes usually increase and decrease acidity, respectively. The processes are catalysed by enzymes from microorganisms and are discussed in Section 5.8. One of the important reactions that causes acidification is:

$$NH_4^+ + 4O \rightarrow 2H^+ + NO_3^- + H_2O, \qquad (9.30)$$

which represents the process of nitrification (Equations 9.28 and 9.29). It occurs with all fertilizers containing ammonium or urea. Thus:

$$(NH_4)_2SO_4 + 8O \rightarrow 2HNO_3 + H_2SO_4 + 2H_2O; \qquad (9.31)$$

$$CO(NH_2)_2 + 2H_2O \rightarrow (NH_4)_2CO_3; \qquad (9.32)$$

and $\qquad (NH_4)_2 CO_3 + 8O \rightarrow 2HNO_3 + H_2CO_3 + 2H_2O. \qquad (9.33)$

The acidifying effect of these fertilizers in soil depends largely on whether crop plants are present, which they normally would be unless the crop failed. In the absence of a crop, the proton equivalent of 28 kg N (2 kg atoms of N) is 4 kg for ammonium sulphate (Equation 9.31) and 2 kg for urea (Equations 9.32 and 9.33), assuming that H_2CO_3 dissociates to CO_2 and H_2O. If the nitrate and sulphate that are produced are leached out of the soil they would carry with them 80 and 40 kg calcium from the addition of the ammonium sulphate and urea, respectively. In the presence of a crop the acidifying effect depends on the extent of nitrification and on the relative uptake of cations and anions (Section 7.6). This makes it difficult to calculate the extent of acidification in the presence of a crop, but it appears generally to be about half of that in its absence, as calculated above.

There are other oxidations:

$$4Fe^{2+} + 2O + 10H_2O \rightleftharpoons 4Fe(OH)_3 + 8H^+; \qquad (9.34)$$

$$S^{2-} + 3O + H_2O \rightleftharpoons H_2SO_4; \qquad (9.35)$$

$$C_6H_{12}O_6 + 2O \rightarrow 2CH_3COCOOH + 2H_2O. \qquad (9.36)$$

Oxidation of iron and sulphide causes acidification, as shown by Equations 9.34 and 9.35. Oxidation of pyrite (a ferrous sulphide) can give very acid soils as described below (see Equation 9.38).

Equation 9.36 is for the oxidation of a sugar, e.g. glucose, to pyruvic acid, which occurs during the aerobic respiration of microorganisms. Oxygen acts as an electron acceptor under these conditions. In anaerobic conditions there are other electron acceptors such as Fe^{3+}, which becomes reduced to Fe^{2+} with the consumption of H^+ (Section 5.8). Hence when acid soils are flooded they become anaerobic and the pH rises.

4. Increase of humus content

Humus has pH-dependent negative charges, which give it a cation exchange capacity at pH 7 of about 3 moles of charge per kilogram; the value differs between soils and depends on the method of measurement. The charge at pH 7 arises mainly from the dissociation of protons from carboxylate groups (Section 2.8). The acid groups are formed by oxidation by microorganisms of plant material which is incorporated in the soil. If the plant material has a high content of basic cations they balance the negative charge on the humus. Measurements in Australia have shown that pastures containing an introduced legume such as subterranean clover can cause acidification. This is probably because there is a shortage of basic cations to balance the acidic groups on the increased content of soil organic matter, although fixation of nitrogen by the clover might be a partial cause.

5. Weathering of minerals

Many primary minerals contain the basic cations Ca^{2+}, Mg^{2+}, Na^+ and K^+ which pass into solution when the minerals hydrolyse, a process of chemical weathering (Section 3.2). The process can be expressed in terms of consumption of H^+ as in Equation 3.1, or production of OH^- expressed generally as

$$(\text{Ca,Mg,Na,K,Al}) \text{ silicate} + H_2O \rightarrow Ca^{2+} + Mg^{2+}$$
$$+ Na^+ + K^+ + Al(OH)_3 + Si(OH)_4 + OH^-. \quad (9.37)$$

The products may include silicic acid, hydrous oxides of aluminium and iron, kaolinite or other clay silicates, and the basic cations that pass into solution. Because of the production of basic cations and OH^- (or consumption of H^+) the process is known as alkalinization. Its effects are greatest when readily hydrolysed minerals, for example ferromagnesian minerals, are present in the soil. The higher the temperature the faster it proceeds. The rate is also increased by addition of acids, but it is not known whether acidification, as it occurs under field conditions, has any effect.

The effect of weathering of most minerals is to increase alkalinity, although oxidation of a ferrous sulphide such as pyrite releases protons, as mentioned above. Pyrite is formed under reducing conditions, especially in marine and estuarine clays and in peats. It is formed by the reduction of sulphate and iron by microorganisms when organic matter is present. It also occurs in the waste material from coal mining. Reclamation of this waste and the drainage of marine and estuarine clays and peats can give very acid soils. They are known as acid sulphate soils and can have a pH below 3. The equation for the oxidation of pyrite is:

$$4FeS_2 + 15O_2 + 8H_2O \rightarrow 2Fe_2O_3 + 16H^+ + 8SO_4^{2-}. \quad (9.38)$$

In ending this section three points may usefully be re-stated.

1. During acidification the concentration of protons in the soil is increased and basic cations are displaced from exchange sites on clays and humus.
2. The basic cations are lost permanently if they are leached out of the soil or removed in a harvested crop.
3. Because of the cycling of basic nutrients and their accumulation in woody structures, as in trees, distinction needs to be made between acidification of soil and that of the ecosystem.

9.6 Effects of soil acidity on plants

Acidity creates chemical and biological conditions in soil which are harmful to many plants, although there are exceptions. The effects are summarized in Table 9.3.

When soils become acid their capacity to adsorb cations is reduced

Table 9.3. *The main effects of soil acidification on plants*

1. Reduced supply of nutrients; Ca^{2+} and other cations lost in drainage water; N, P and S remain immobilized for longer in the soil organic matter.
2. Increased concentrations of metal ions in solution, especially of Al, and including those of Mn, Cr, Cu, Ni, Zn, which may become toxic.
3. The uptake of phosphate and molybdate is reduced.
4. The form of nitrogen taken up by plant roots may be NH_4^+ instead of NO_3^- because nitrification is inhibited.
5. Nitrogen fixation by legumes may be reduced unless the *Rhizobium* strain is acid-tolerant.

(see 'effective cation exchange capacity' in Section 4.2) so that nutrient cations, especially Ca^{2+} and Mg^{2+}, pass into solution and are leached in drainage water. Acidity also increases the solubility of most metals. As the pH decreases below about 5.5 soil solutions contain increasing concentrations of aluminium ions; $Al(OH)_2^+$, $AlOH^{2+}$ and Al^{3+} become dominant in that order as the pH falls, and they displace other cations from exchange sites. Acid soils therefore usually have low contents of calcium and magnesium, and in extreme conditions the supply to plants may be deficient.

The most common problem in acid soils is, however, the toxicity of aluminium to plants, and for some species the toxicity of manganese. Solution concentrations of 10^{-6} M Al can harm sensitive plants. The concentrations of other metals, including chromium, copper, nickel and zinc are increased in acid soils, but their toxicity is almost entirely restricted to polluted soils (see Chapter 10).

Acid soils can affect plants in several ways (Table 9.3). Plant species and varieties differ, however, in their sensitivity to the conditions in acid soils. Among crop plants, barley, cotton and lucerne (alfalfa) are sensitive whereas tea actually requires acid conditions. The usual agricultural practice for most crops, at least in temperate regions, is to maintain a soil pH of 6.0–6.5 by the addition of lime, applied as calcium carbonate, calcium hydroxide or calcium oxide. The selection and breeding of varieties of crop and pasture plants less sensitive to acid conditions lessens the need for lime.

9.7 Acid rain

The term 'acid rain' was first used in the nineteenth century to describe rain in the industrialized parts of north-west England that contained

Table 9.4. *Soil acidification: atmospheric sources*

Sulphuric acid	Main source: atmospheric oxidation of SO_2 (see below) and reduced sulphur compounds.
Nitric acid	Main source: atmospheric oxidation of oxides of nitrogen (NO_2, NO, N_2O); see below.
Hydrochloric acid	A combustion product of coal.
Carbonic acid	Formed by CO_2 dissolving in water.
Sulphur dioxide	Man's activity: coal burning (main source), burning of other fossil fuels, smelting of metal sulphides. Natural source: volcanoes.
Reduced sulphur compounds	H_2S, released on combustion of crude oil, sulphur mining and some industrial processes; a product of microbial reduction. Dimethyl sulphide (DMS), $(CH_3)_2S$, dimethyl disulphide, $(CH_3S)_2$, and carbon disulphide, CS_2, products of reduction; DMS is of marine origin.
Oxides of nitrogen	Nitrous oxide, N_2O, and nitric oxide, NO, products of microbial reduction. Nitric oxide and nitrogen dioxide, products of combustion of fossil fuels.
Ammonia	Released from soils with pH > 7 and from organic manures; soil acidity caused by microbial oxidation to nitrate (nitrification).

acid pollutants. Later, it became known that the deposition of certain gases and salts from the atmosphere also cause acidification. The latter are known as dry deposition. 'Acid rain', although a term that is in common use, is therefore an incomplete description of the total acidifying effects of deposits from the atmosphere. The main acids and acid-forming chemicals deposited on the Earth's surface are listed in Table 9.4.

Sources of the acids

The acids that gave rise to the term 'acid rain' were sulphuric, nitric and hydrochloric. Their main source is the combustion of fossil fuels; of the three, sulphuric and nitric acids are the most important.

The combustion of fossil fuels, especially of coal and, to a less extent of oil, in power stations, and the smelting of metal ores containing sulphide, produces sulphur dioxide, SO_2, which is emitted into the atmosphere (Table 9.5). In the United Kingdom the emissions increased throughout the nineteenth century, they reached a peak in 1970 of 3 million tonnes of SO_2-S and have since decreased by 40%. Some of the

Table 9.5. *Estimates of global emissions to the atmosphere of gaseous sulphur compounds*

Source	Annual flux of S (Mt)
Anthropogenic (mainly sulphur dioxide from fossil fuel combustion)	80
Biomass burning (sulphur dioxide)	7
Oceans (dimethyl sulphide)	40
Soils and plants (hydrogen sulphide and dimethyl sulphide)	10
Volcanoes (hydrogen sulphide and sulphur dioxide)	10
Total	147

Source: From Houghton, J.T., Jenkins, G.J. and Ephraums, J.J. (eds) 1990, *Climate Change, Intergovernmental Panel on Climate Change.* Cambridge University Press; with permission.

SO_2 may be deposited on plants and soil, some may dissolve in water droplets in the atmosphere, and some is oxidized by complex reactions in the atmosphere that probably involve highly reactive hydroxyl radicals (see Section 11.3). The product is sulphur trioxide, SO_3, which dissolves in water droplets to form sulphuric acid.

Fossil fuels also contain nitrogen compounds, power stations and motor vehicles being the main sources of oxides of nitrogen emitted from these fuels. The products of combustion depend on temperature. In the high temperature processes of power stations and in the internal combustion engine the main product is nitric oxide, NO. In the atmosphere nitric oxide is oxidized to nitrogen dioxide, NO_2, and further reaction with ozone and hydroxyl radicals produces nitric acid, HNO_3 (Equations 11.7 and 11.8).

The sources of ammonia also need to be mentioned. These are mainly from agricultural activities. Ammonia is volatilized from soils of pH greater than 7.0, as commonly occurs in soil containing calcium carbonate, and also after application of urea as fertilizer (Section 8.5). Organic manures release ammonia, high concentrations being present in the atmosphere close to intensive animal units. Much of the ammonia is absorbed locally by soil, plants and water, but some passes into the atmosphere where it dissolves in rainwater and may also form aerosols of ammonium sulphate and ammonium nitrate.

Acidifying deposits from the atmosphere

As mentioned in Section 9.5, rainwater in equilibrium with the atmosphere and containing carbonic acid as the only dissolved acid has a pH of 5.6. In a clean environment it may contain small concentrations of sulphuric, nitric and hydrochloric acids of natural origin and have a pH of about 5.0. In urban and industrial areas, where rainwater has a higher concentration of these acids, especially sulphuric and nitric acids, the pH may be about 4.0, or less.

The rate of deposition of hydrogen ions can be calculated from the amount of rainfall and its chemical composition. In the United Kingdom the annual deposition is commonly in the range 0.1–0.6 kg H^+ per hectare in rural areas, and about 1–1.5 kg H^+ per hectare near urban and industrial centres. The average annual deposition at 16 rural sites in the 1980s was 0.3 kg H^+ ha^{-1} (Table 9.6). The composition of rainwater is influenced by marine salts; when allowance is made for these the acidity can generally be attributed to sulphuric and nitric acids. The annual deposition of these acids is greatest near urban and industrial centres and least in remote rural areas. In the United Kingdom the annual deposition in rainfall is commonly in the range 5–10 kg S ha^{-1} as non-marine sulphate, and 2–5 kg NO_3^-—N ha^{-1}.

In addition to these acids, ammonium ions are also present in rainwater. In rural areas the average annual deposition of NH_4^+—N given in Table 9.6 is 4 kg ha^{-1}; most values in the United Kingdom are between 2 and 8 kg ha^{-1}. Oxidation of NH_4^+ produces 2 moles of H^+ per mole of NH_4^+ (Equation 9.30), so 4 kg of NH_4^+—N will produce about 0.6 kg H^+ per hectare (Table 9.7).

Dry deposition includes the absorption of sulphur dioxide, nitrogen oxides, nitric acid and ammonia and the retention of aerosols, which may contain sulphuric acid, nitric acid and ammonium salts. Their total contribution to acidification is not yet known, but there is information on the deposition of sulphur dioxide and ammonia.

In the United Kingdom the annual dry deposition of sulphur, mainly as sulphur dioxide, varies greatly from below 3 to over 30 kg S ha^{-1} depending on the distance from its source. Oxidation in soil of 20 kg SO_2—S ha^{-1} produces 1.3 kg H^+ per hectare according to Equation 9.25, although this may be an overestimate. Although there have been few measurements of the dry deposition of ammonia, the usual range in the United Kingdom is probably 10 to 30 kg NH_3—N per hectare per year. An annual rate of deposition of 20 kg NH_3—N per hectare will produce

Table 9.6. *Mean annual deposition in rainfall at 16 rural sites in the United Kingdom, 1981–5*

Figures are kilograms of element per hectare.

	H^+	SO_4^{2-}-S	NO_3^--N	NH_4^+-N	Na^+	Cl^-	Mg^{2+}	Ca^{2+}
Range	0.09–0.61	3.9–18.9	1.5–5.1	2.7–8.1	5.7–93.3	13.5–172.0	1.0–11.6	3.5–8.7
Mean	0.31	8.6	3.1	4.0	34.1	61.0	4.5	5.3

Source: From *Acid Deposition in the United Kingdom* 1981–1985, DOE 1987; Crown copyright.

Table 9.7. *Potential soil acidification from atmospheric deposits commonly found in the United Kingdom*

Source	Deposition[a] (kg ha^{-1} a^{-1})	Proton equivalence[b] (kg ha^{-1} a^{-1})
Rainfall		
H$^+$	0.3	0.3
NH$_4^+$-N	4.0	0.6
Dry deposition		
SO$_2$-S	20	1.3
NH$_3$-N	20	1.4

[a]Deposition in rainfall from Table 9.6; for dry deposition see text.
[b]The proton equivalence is calculated assuming that NH$_4^+$, SO$_2$ and NH$_3$ are oxidized completely in soil; absorption by vegetation will reduce the proton equivalence.

1.4 kg H$^+$ on oxidation (see Equation 9.27), although again this may be an overestimate.

To summarize, deposits from the atmosphere which cause soil acidification include acids in rainwater, but ammonium ions in rainwater and the dry deposition of sulphur dioxide and ammonia will generally have greater effects, as shown in Table 9.7. The values of proton equivalence may overestimate the immediate acidification of soils because of the absorption by vegetation.

The effect of acid inputs will differ between soils according to their capacity to neutralize acids and buffer the pH change (Section 9.3). Maps are now being produced to show the critical load of acid input that soils can withstand without incurring long-term damage.

The environmental effects of acidification are described in Section 9.9.

9.8 Acidification of ecosystems

In this section we examine the evidence for soil acidification in natural and agricultural ecosystems. This will first be done for the loss of calcium carbonate and pH decrease, and then examples will be given of the relative importance of the various processes of acidification.

Table 9.8. *Postulated annual loss of calcium carbonate from soils of pH 5.0–8.0 with 250 mm of through-drainage*

Soil pH	Calcium carbonate loss ($kg\ ha^{-1}\ a^{-1}$)
8.0	942
7.5	672
7.0	471
6.5	336
6.0	235
5.5	168
5.0	118

Source: From Gasser, J.K.R. 1973. *Experimental Husbandry*, **25**, 86–95.

Loss of free carbonates and calcium

The free carbonates in soil are $CaCO_3$ and $MgCO_3$, of which $CaCO_3$ is quantitatively the more important in agricultural and natural ecosystems and will be used as the example. It has low solubility in CO_2-free pure water, but in the presence of carbonic acid, the soluble bicarbonate is formed (Equation 9.14). The solubility depends on the partial pressure (or concentration) of CO_2 in the air with which the solution is in equilibrium. In soil, where the CO_2 concentration of the air is higher than in the atmosphere, the dissolution of calcium carbonate gives a concentration of 0.001 to 0.01 M Ca^{2+} and a pH of between 7 and 8.4.

When rainwater passes through the soil, calcium and other cations are leached with bicarbonate ions and also with sulphate and nitrate ions when these are present in the rainwater or are formed in the soil. The amounts of calcium that can be leached by 250 mm of drainage water containing CO_2 in equilibrium with the atmosphere are given in Table 9.8.

Because of the dissociation of carbonic acid (Equation 9.13) rainwater passing through soil contains protons which displace the exchangeable basic cations. The dissociation decreases to almost zero at pH 5. Other acids (Section 9.5) increase the hydrogen ion concentration.

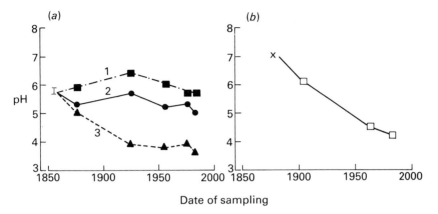

Date of sampling

Figure 9.3 Acidification of soil at Rothamsted Experimental Station, (*a*) under grass and receiving fertilizers, and (*b*) unmanured and allowed to revert to woodland. The pH is of surface soil samples (0–23 cm). In (*a*) plot 1 received complete fertilizer with N supplied as nitrate, plot 3 received complete fertilizer with N supplied as ammonium, and plot 2 was unmanured. (From Goulding, K.W.T. and Johnson, A.E. In *Acid Deposition* (ed. J.W.S. Longhurst). British Library; Crown copyright.)

Examples of pH change

Because of the buffer properties of soil, decrease of pH alone cannot be used to quantify the effects of additions of acids and acid-forming chemicals. Soil pH is, however, the only measurement that is usually made, and this can change rapidly; Figure 9.3 shows two examples.

The Park Grass plots at Rothamsted Experimental Station have been under permanent grass. The pH of unfertilized plots decreased from 5.7 to 5.1 in 130 years whereas over the same period the pH decreased to 3.6 with annual addition of 96 kg ha^{-1} of N as ammonium sulphate (Figure 9.3*a*). Applying Equation 9.31, oxidation of 28 kg NH_4-N ha^{-1} produces 4 kg H$^+$, and 96 kg N therefore produces 14 kg H$^+$ ha^{-1}, although the annual production of acid was probably not as high.

At the Geescroft Wilderness site, also at Rothamsted, arable land was allowed to revert to woodland in 1890 and then received no treatment. Over a period of about 100 years the pH of the surface soil (0–23 cm) decreased from about 7 to 4.2 (Figure 9.3*b*). The processes cannot be quantified with certainty, but a contributing process is probably the capturing of acids from the atmosphere (wet and dry deposition) by the trees. This is in sharp contrast with the neighbouring unfertilized Park Grass plots referred to above.

9.9 Environmental effects

The effects of acids and acid-forming chemicals in the atmosphere on terrestrial and freshwater systems have been attributed to (1) the enrichment of nitrogen and sulphur supplies (eutrophication), (2) forest decline, that is, loss of needles and leaves, die-back of the crown of the tree and death in extreme cases, and (3) decline of populations of fish in lakes and streams; each will now be considered.

Eutrophication

Deposition of nitrogen and sulphur compounds from the atmosphere provides essential nutrients. In the United Kingdom the annual rate of deposition from non-marine sources is commonly in the range 8–40 kg S ha^{-1} and 20–30 kg N ha^{-1}. These additional nutrients, and especially nitrogen, affect the distribution of species in natural habitats and are believed to increase tree growth and the yields of agricultural crops. There will be harmful effects if luxury uptake of nitrogen leads to disease or increased damage by insects.

Forest decline

Forest decline, known as 'Waldsterben' (forest death) in Germany where the problem has received much attention, has been attributed to the effects of acid rain. In parts of Germany, northern Czechoslovakia and southern Poland damage and death of conifers such as silver fir (*Abies alba*) has been attributed to high concentrations of atmospheric SO$_2$. There has been concern that trees have become unhealthy or died in parts of several other countries including Sweden, Norway, the United Kingdom and the United States.

In the most severely affected parts of eastern Europe the cause appears to be atmospheric pollution, especially by SO$_2$. Elsewhere, the causal agent, or agents, probably differ between sites. One difficulty in less affected areas is the shortage of reliable information from surveys to establish the extent of forest decline. Another is that good experimentation on the effects of atmospheric pollutants on trees has been undertaken only recently and the outcome is not yet clear.

Forest decline in areas with relatively low concentrations of atmospheric SO$_2$ has been attributed to several causes. Those commonly reported include:

(i) Accumulation of aluminium and deficiency of magnesium (possibly also of calcium) in tissues of old trees growing on acid soils.

(ii) Increased damage by insects and disease organisms because increased uptake of nitrogen has made plant and tree tissues softer (see above).

(iii) Ozone, which has increased in recent years in the lower part of the stratosphere, itself damages plants and trees. It has also been shown to make Norway spruce (*Picea abies*) more susceptible to frost injury, and in combination with acid mists has led to increased needle fall by Scots pine (*Pinus sylvestris*) and leaf fall by beech (*Fagus sylvatica*).

(iv) Adverse weather conditions including hot dry summers in the 1970s and 1980s, and cold winters.

(v) Poor forest management.

There may also be synergism between these various causes.

Decline of fish populations

The disappearance of brown trout (*Salmo trutta*) from mountain lakes in Scandinavia began in the 1900s and about 50 years later was associated with the acidification of the water. The problem has also been recognized in parts of North America and the United Kingdom. Numbers of other fish species including the Atlantic salmon (*Salmo salar*) have also declined in these areas. The increasing acidity of small lakes in south-west Scotland has been shown by change of species of diatoms, which differ in sensitivity to pH. The decrease of pH in one lake, estimated from the change in species of diatoms, was from 5.6 in the middle of the nineteenth century to 4.7 by 1980. During this period there were no significant changes of vegetation or land use, and acidification has therefore been attributed to acid deposition from the atmosphere.

The damage to fish is due less to acidity itself than to aluminium in acidic solutions. Laboratory studies have shown that survival of the fry of brown trout is reduced by concentrations of aluminium of 250 $\mu g\ l^{-1}$ and their growth is reduced by one tenth of this concentration; these concentrations can occur in acid waters. The effect of aluminium becomes less if the concentration of calcium is increased, and silica may have a similar effect. At high concentrations in solution aluminium accumulates in the gills of fish, so stopping oxygen transfer. It is now generally accepted that the damage to fish is associated with low pH, high aluminium and low calcium concentrations.

The high aluminium concentration is caused by acidic solutions passing through soil and weathering rock. Aluminium passes into solution as Al^{3+} and $Al(OH)^{2+}$, which are toxic to fish, and it may be present in less toxic organic complexes. The lakes and rivers most at risk are those in areas where there is acid deposition from the atmosphere on thin soils over low-base rocks such as granite. Adding powdered calcium carbonate to the water or the surrounding catchment area is the treatment usually recommended when acidity increases above a critical level.

9.10 Summary

Acidification of soils is a natural process which is accelerated by deposition of increased amounts of acids and acid-forming substances from the atmosphere ('acid rain'). Acidic conditions can have harmful effects on plants and trees, and aluminium leached from acid soils can be toxic to fish.

There are several causes of acidification: oxidation of NH_3, NH_4^+, S^{2-} and Fe^{2+}; increase of humus content; secretion of protons by plant roots and removal of basic cations in crop harvests; deposition of acids (H_2CO_3, HNO_3, H_2SO_4, HCl), acid gases (CO_2, SO_2, SO_3, HNO_3) and acid-forming substances (NH_3 and NH_4^+).

The soil buffers pH change by various reactions between protons and calcium and magnesium carbonates, clay minerals and humus. The weathering of most minerals (sulphides are an exception) leads to alkalinization.

10

Heavy metals and radionuclides in soil

10.1 Introduction

There are two parts to this chapter. Sections 10.2–10.6 deal with elements generally associated with industrial and urban pollution, most of which are metals that are often described as heavy metals (see Section 10.2). Sections 10.7–10.9 are more specific and describe some of the environmental problems caused by radionuclides (also called radioactive nuclides or radioactive isotopes which decay with time, in contrast to stable isotopes).

Heavy metals are natural components of the environment, but are of concern because they are being added to soil, water and air in increasing amounts. Some, for example copper, manganese and zinc, are micronutrients (trace elements) which are essential in small amounts for plant and animal life (Section 7.5). They can, however, be harmful if they are taken up by plants or animals in large amounts, as can other heavy metals not known to be essential nutrients. The objective of research is to establish the amounts, or concentrations, that are safe. It should be noted that advances in instrumentation have made possible the measurement of concentrations which were previously not detectable, but it does not follow that these low concentrations necessarily indicate any hazard to health.

Radionuclides are a danger, rarely from the elements themselves but from the radiation they emit (plutonium is an exception, being highly toxic to animals). Some occur naturally, exposing all living organisms to small and usually harmless doses of radiation. Particularly dangerous are some man-made radionuclides that accumulate as they pass through a biological cycle. For example, the chemistry of strontium is similar to that of calcium, and radionuclides of strontium accumulate with calcium in bone tissue, where they emit radiation.

The principal pathways by which the heavy metals and radionuclides enter the soil–plant continuum in amounts that might present a hazard are deposition from the atmosphere, seepage from waste disposal sites, and application of sewage sludge, pig manure and certain fertilizers and pesticides; amounts can also be high close to bodies of metal ores. Entry into the plant is normally in ionic form, as for the essential elements (Section 7.5). The chemistry of their reactions in soils and plants, however, shows little uniformity, which is to be expected considering the wide range of elements to be discussed. The unifying theme is that of environmental, and particularly soil, contamination.

It should be noted that organic substances can also contaminate soil. They include pesticides, discussed in Section 13.4, and also engine oil and several waste products from industry which will not be discussed.

10.2 Heavy metals: definition

The term 'heavy metals', which is in common use, refers to metals with a density greater than a certain value, usually 5 or 6 g cm^{-3}. Often it refers to metals discharged by industry of which the metalloid arsenic, As; cadmium, Cd; chromium, Cr; copper, Cu; lead, Pb; mercury, Hg; nickel, Ni; and zinc, Zn, are listed by a European Commission directive as representing the greatest hazard to plants or animals. As used here, the term refers to hazardous metals, usually of high density, whatever their source; 'hazardous elements' is used where both metals and non-metals are to be described.

All foods and water contain metals and non-metals that at high concentrations, whether from natural or industrial sources, can become harmful. Some are particularly hazardous and it is these that will be considered in this chapter.

10.3 Hazardous elements in soils

The concentrations in the Earth's crust and soils of elements considered to be a hazard are given in Table 10.1. Concentrations differ greatly between rock types, and because soil properties are influenced by their parent material, concentrations in soils also vary greatly. In igneous rocks the elements are incorporated in the structure of minerals when the magma crystallizes. The concentration in sedimentary rocks is determined partly by the minerals they contain and partly by adsorption from the water in which the sediment is deposited.

Soils also receive chemical elements from active volcanoes, including arsenic, fluorine, mercury and selenium, and these can be a hazard. Of greatest concern, however, are emissions of metals into the environment as a result of man's activities. Emissions from industry and from domestic waste disposal are the biggest of the anthropogenic sources (Table 10.2). The bulk of the metals are disposed of as solid wastes in landfill sites.

The effects of emissions from industry depend on their nature. Solid waste from the mining of metal ores has a local effect whereas emissions into the atmosphere from metal refineries can contaminate large areas

Table 10.1. *Average concentrations of some hazardous metals*

Metal	Earth's crust ($\mu g\ g^{-1}$)	Rocks with highest concentration	Soils ($\mu g\ g^{-1}$)	Soils[a] (kg ha^{-1})
As	1.5	shales and clays	0.1–50	0.2–100
Cd	0.1	shales and clays	0.01–2.4	0.02–4.8
Cr	100	ultrabasic	5–1500	10–3000
Cu	50	basic	2–250	4–500
Hg	0.05	sandstones	0.01–0.3	0.02–0.6
Ni	80	ultrabasic	2–1000	4–2000
Pb	14	granite	2–300	4–600
Zn	75	shales and clays	10–300	20–600

[a]Amount of metal per hectare calculated for a soil depth of 15 cm and a bulk density of 1.3 (approximate mass 2000 t).
Source: Adapted from Alloway, B.J. (ed.) 1990. *Heavy Metals in Soils.* Blackie, Glasgow.

Table 10.2. *Annual release of metals into the environment in the United Kingdom*

	Cd	Cu	Ni	Pb	Zn
Total release (t a^{-1})	754	15 800	14 800	51 700	85 000
Contribution to total release (%)					
from industry	25	23	27	33	31
municipal waste disposal	67	65	71	65	64
domestic	1	4	<1	<1	1
urban run-off	<1	<1	<1	<1	<1
agriculture	5	5	<1	<1	1
dredging spoil	1	3	1	1	2
Total release received in landfill sites (%)	83	84	91	81	88

Source: data of Critchley and Agg, quoted by Beckett, P.H.T. 1989. In *Inorganic Contaminants in the Vadose Zone* (eds B. Bar-Yosef, N.J. Barrow and J. Goldschmidt), pp. 159–75, Springer-Verlag, Berlin; with permission.

of land. Similarly, the metals in sewage sludge are more widely distributed on agricultural land than those disposed of in landfill sites. These sources will now be discussed.

Mining and smelting

Mining operations usually entail the tipping of low-grade ore and other solid waste, and some may accumulate in river alluvium. Fine particles of the tailings, or of the ore, can be blown or washed onto adjacent land carrying with them high concentrations of the metal being mined and also associated metals. One example is that soils near old lead mines in Wales contain up to about 10 000 µg Pb g^{-1} soil, more than ten times the top of the range given in Table 10.1.

During smelting to purify the metal ore, particles and gases, especially sulphur dioxide, are emitted. Particles from the smoke are deposited at a rate dependent on their size and density; very fine particles (aerosols) and gaseous products are washed out by rain. Deposition of metals is greatest near the smelter and decreases exponentially with distance. The prevailing wind determines the direction in which the plume of particles and gases is carried. One of the best known examples of contamination is from the copper–nickel smelter at

Sudbury, Ontario, Canada, where soils within 7.5 km of the smelter contained over 1000 μg Cu g^{-1}. When soils were analysed on a transect from a smelter at Avonmouth, Bristol in southwest England, background levels of lead and cadmium were not reached at 14 km.

Landfill sites

Particular attention has to be given to landfill sites because of their use for the disposal of large amounts of hazardous elements (Table 10.2). Precautions have to be taken to prevent drainage and overflow water from the sites containing hazardous metals or other substances.

The sites chosen for landfill in many countries require planning permission. They are usually holes left after quarrying for stone or extraction of gravel. Ideally they should have a base of clay of low water permeability. A normal requirement is for the site to be sufficiently far from rivers and groundwater to prevent contamination from drainage and overflow liquids, both of which should have their composition monitored. An impermeable cover may be needed to stop the infiltration of rainwater. Planning permission carries with it a requirement that when filling has been completed the land should be restored to agriculture, forestry or amenity use by the return of topsoil. Building on the land is not recommended.

The materials put into landfill sites are municipal and industrial waste. The composition varies, but more than half of municipal waste may be paper, with the rest made up of food scraps, metals, glass, and ash. Wastes from industrial plant may contain acids, alkalis, oils, metals and other noxious substances; they need special care and may need to be contained within the site by plastic liners. Municipal waste accounts for the bulk of the material put into landfill sites.

Because of the presence of large amounts of decomposable organic substances collected by municipal authorities intense microbial activity occurs in the buried waste. From being initially aerobic the waste becomes anaerobic, a condition that can last for several years. Reduction by microorganisms produces methane and can also produce methylated compounds, for example of arsenic and mercury, which are volatile. There are, however, no known cases of toxicity arising from these or any other metal in volatile compounds at landfill sites. Reduction (see Section 5.8) also occurs of manganese and iron oxides to Mn^{2+} and Fe^{2+}, respectively, both of which are soluble and liable to be leached. Microbial decomposition results in loss of organic materials (to

CO_2 and CH_4 mainly) which causes settlement of the site of up to about 25% of the depth of the landfill. The mineral components remain as a potential hazard to water supplies unless the site has been properly sealed. The methane which is generated can be used as an energy source, but can be a hazard if allowed to escape because it forms an explosive mixture with air. (See also Chapter 11 for its chemical effects in the atmosphere.)

Sewage sludge

Sewage sludge is the organic material produced from domestic and industrial waste water and direct run-off from roads. In the United Kingdom about 90% of the material is treated at sewage works, most of the rest at present (1991) being discharged without treatment into the sea. The primary treatment is settlement for about 24 hours, to remove coarse particles. The supernatant liquid receives a secondary treatment by being sprinkled once and sometimes twice through biologically active filters. Further filtration or flocculation (tertiary treatment) may follow before the liquid is discharged into rivers. Sludge which is to be applied to the land is usually digested anaerobically to reduce its water content and smell, and further dewatered mechanically or by allowing it to dry. In the United Kingdom 40 million tonnes of sewage are produced annually, containing 1.3 million tonnes of dry matter. About 40% is applied to agricultural land, and using the analysis in Table 10.3 this will add about 12, 6 and 3 kilotonnes of N, P and K respectively.

The composition of sewage sludge is very variable. It depends on the local industrial processes and on the amount of sand and silt that it contains. It is useful as a source of nitrogen and phosphate for plants, but has only a small content of potassium because most remains in the liquid that is discharged into rivers.

The organic matter in sewage sludge helps to improve soil structure. When applied to land it therefore has beneficial effects. Problems arise if large amounts are applied too frequently or over a prolonged period and the sludge contains high concentrations of metals which are toxic to plants or animals. Those considered to present the greatest hazard are shown in Table 10.4. Concentrations vary greatly according to the source of the waste, and particularly on the amount of industrialization and nature of the industrial processes in the catchment area. As examples, chromium and nickel are released by the iron and steel industry, cadmium and lead from the manufacture of batteries, and zinc from zinc

Table 10.3. *Typical analysis of sewage sludge (% moist sludge)*

	Moisture	N	P	K
Sludge cake	45	1.2	0.6	0.3
Annual application to agricultural land in UK (kt)	—	12	6	3

Table 10.4. *Concentrations of heavy metals (mg kg^{-1} dry matter) in 42 sewage sludges from England and Wales*

Metal	Median	Range	kg in 25 t
Cd	—	<60–1500	—
Cr	250	40–8800	6.3
Cu	800	200–8000	20.0
Ni	80	20–5300	2.0
Pb	700	120–3000	17.5
Zn	3000	700–49 000	75.0

Source: From Berrow, M.L. and Webber, J. 1972. *Journal of Science of Food and Agriculture* **23**, 93–100.

plating factories. Zinc and copper predominate in domestic sewage but are usually present in lower concentrations than in sewage that contains waste from industrial processes.

The contamination of sewage sludge by organic chemicals of industrial, including agricultural, origin and also by pathogens also needs to be mentioned but is outside the scope of this book.

Two systems have been used to define the permissible upper limit for the addition of metals in sewage. The first was to recommend a maximum addition of 250 μg 'Zn equivalent' per g soil, based on the composition of the sewage sludge as:

$$\text{'Zn equivalent'} = (1 \times \text{Zn}) + (2 \times \text{Cu}) + (8 \times \text{Ni}).$$

The concentration of each component is expressed in micrograms per gram of dry matter of sewage. The formula is based on the relative phytotoxities of the three elements and on the assumption that their effects are additive. The maximum addition of 250 μg 'Zn equivalent' applies to a soil of pH not less than 6.5; the threshold value is less for acid soils because of the greater solubility of the metals. Although the

Table 10.5. *Maximum metal concentrations in soils permitted under*
European Community regulations

Metal	Maximum soil concentration (mg kg^{-1})	Metal	Maximum soil concentration (mg kg^{-1})
Zn	300	Cd	3
Cu	140	Pb	300
Ni	75	Hg	1.5

'Zn equivalent' has been superseded, it provides a useful guide for sewage additions as the following example shows.

For an application of 25 t (25 × 10^6 g) per hectare and using the median concentrations shown in Table 10.4, the addition of the 'Zn equivalent' is

$$[(1 \times 3000) + (2 \times 800) + (8 \times 80)] \times 25 \times 10^6/(2 \times 10^9)$$

$$= 65.5 \text{ µg per g soil.}$$

This assumes a soil bulk density of 1.3 so that the mass of 1 hectare of soil to 15 cm depth is 2000 t (2 × 10^9 g). If four applications, each of 25 t dry matter are made, the threshold of 250 µg g^{-1} soil would be slightly exceeded.

The 'Zn equivalent' takes into account only the metals present in sewage. There is now a requirement of a European Commission that the concentration of each potentially toxic metal in soil should not exceed a defined threshold as shown in Table 10.5. (This requirement also applies to potentially toxic non-metals such as fluoride.) Contamination from sources other than sewage and naturally high metal concentrations in soil are therefore taken into account. A requirement from 1 January 1992 is for soils to be sampled and analysed if they are to receive sewage.

When guidelines have been followed applications of sewage sludge have rarely caused harm to crops. Most reports of phytotoxicity are from sewage farms where the sewage has been applied in large amounts for several years. Red beet and other vegetable crops have been found to be sensitive to high concentrations of metals. There can, however, be a long-term problem because most of the metals remain in the soil where they might represent a hazard. Such a hazard is illustrated by observations at the Woburn farm of Rothamsted Experimental Station where sewage sludge had been applied to a sandy loam soil of pH 6.5 at an annual rate of 16.4 t of organic matter between 1942 and 1967. White clover (*Trifolium repens*) grew poorly on the soil 20 years after the last application of sewage. The total mass of microbes in soil was less after treatment with the sewage than after treatment with fertilizer. It was also shown that metals in the sewage inhibited nitrogen fixation by *Rhizobium* in the nodules formed on clover roots.

10.4 Accumulation in plants

As described in Section 7.5, the rate of uptake of nutrient ions by plant roots depends largely on their concentration in the soil soluton at the root surface and on their replenishment in solution. The same applies to ions that are not essential nutrients. The solution concentration is determined by the amount adsorbed by reactive soil particles: humus, oxides of iron, manganese and aluminium, and the clay aluminosilicates. The heavy metals occur in solution as cations and are adsorbed by the negatively charged soil particles. Additionally, however, they are more strongly held as complexes on the surfaces of clay aluminosilicates, hydrated oxides and humus, as discussed in Section 4.5. In general, adsorption increases with pH (molybdate is an exception) so that desorption and the solution concentration are greatest in acid soils. Uptake by plants follows the same pattern (Figure 8.5). Desorption also depends on the activity of microorganisms, which change the pH at microsites and form soluble organic complexes, and on proton release by roots, the effect being greatest in the rhizosphere.

Two further aspects of accumulation in plants need to be discussed. First, some plants are tolerant of high metal concentrations and are able to grow on metalliferous sites, whereas other non-tolerant plants will not grow. A second and related question is the extent to which the elements are transported to the parts of plants which are ingested by humans and animals. Metal tolerance will be discussed first.

The main mechanisms of metal tolerance are as follows.

(i) Exclusion from the roots It has been observed that the roots of heather, *Calluna vulgaris*, contained less copper and zinc when mycorrhizal than when non-mycorrhizal, and similar observations have been made with some tree species. Mycorrhizas can have the opposite effect, increasing metal uptake, and this is probably more common. Exclusion is more usually due to the competitive effect of other cations, including calcium and magnesium, and also other heavy metal cations.

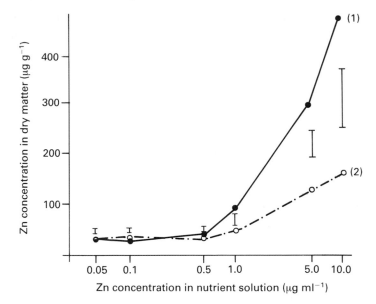

Figure 10.1 The concentration of zinc in the shoots of *Silene maritima* from plants grown in solutions containing a range of zinc concentrations. The plants from a coastal population (1) contained higher concentrations than those from near a zinc mine, (2). The vertical bars are least significant differences (p=0.05). (From Baker, A.J.M. 1978. *New Phytologist* **80,** 635; with permission.)

(ii) Immobilization in roots In many plant species there is restricted translocation of metals, including cadmium, copper, lead and zinc, from roots to shoots. Concentration in the shoots is often less in plants that show tolerance to high metal concentrations in the external solution than those that are sensitive. This difference has been shown for zinc with populations of *Silene maritima* (Fig. 10.1). The metals are retained by cell walls in the roots.

(iii) Biochemical immobilization Plants which accumulate large amounts of metals that are normally phytotoxic, e.g. nickel, form complexes with organic acids which reduce the interference with metabolic processes. Accumulator plants are used to locate deposits of metal ores. Because of their high metal content the shoots of these plants can be a health hazard if eaten.

Metals that are retained by the cell walls of fibrous roots are not ingested by animals and humans. Most of the metals that are trans-

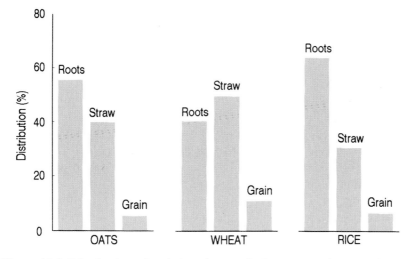

Figure 10.2 Distribution of cadmium in cereal plants grown in pots of soil to which 200 μg Cd as $CdCl_2$ had been added. The distribution in roots, straw and grain is expressed as percent of the total cadmium in the plants. (From Williams, C.H. and David, D.J. 1973. *Australian Journal of Soil Research* **11**, 43–56.)

ported to the shoots accumulate in the leaves, only a small proportion passing to storage organs, which is illustrated in Figure 10.2 for the distribution of cadmium in oats, wheat and rice. As the storage organs form a substantial part of the human diet this provides further protection against toxicity problems.

10.5 Four hazardous elements: Cd, Pb, Zn, F

Several elements have caused fatalities of animals and humans due to the ingestion of excessive amounts, although very few of these fatalities have been caused by contamination of soils, and hence of food, by metals of anthropogenic origin. The four elements to be discussed include fluorine, which is not a heavy metal. They cover a wide range of chemical properties and react with soils in various ways.

Cadmium

This is a relatively rare metal, as shown by its low concentration in the Earth's crust and in soils (Table 10.1). It is present in zinc sulphide ores, and in lesser concentration in zinc carbonates and silicates and in lead

and copper sulphides. It is recovered as a byproduct during the proces-
sing of zinc ores, but its removal is often incomplete. Its increasing
industrial use in batteries, alloys and pigments, as a stabilizing agent for
polyvinyl plastics, and in electroplating metals, has caused concern
about its effects in the environment.

Part of the concern was based on the death of 65 women in Japan that
was reported in the 1950s, for which a causative factor was the high
concentration of cadmium in locally grown rice. The condition is known
as itai-itai disease. The concentration in the rice was high (about ten
times that from an unpolluted area) because the crop was irrigated from
a river contaminated by drainage water and sediment from a zinc mine.
It is the only instance of the poisoning of humans by cadmium which is
attributable to its presence in soils or water.

One of the most widespread sources of cadmium in soils is phosphatic
fertilizer, which has an average concentration of about 7 μg Cd g^{-1},
although this varies with the source of the phosphate rock from which
the fertilizer is manufactured. In one experiment in Australia, the cadmium content of soil was three times as high on plots that had received
1000–4500 kg ha^{-1} of superphosphate over 30–45 years as on neighbouring, unfertilized plots; the uptake by plants was also higher. Generally,
however, much more cadmium is added in one application of sewage
sludge than in a normal application of fertilizer.

In aqueous solution, cadmium is only weakly hydrolysed and the
predominant ion is Cd^{2+}. At the low concentrations found in soil solutions it is adsorbed onto clay minerals including iron, aluminium and
manganese oxides, adsorption increasing with pH. It is also strongly
adsorbed onto calcium carbonate. Organic matter adsorbs cadmium,
but less strongly than copper or lead. Cadmium is, however, more
mobile than the other metals considered in this section and can become
a hazard if washed into domestic water supplies.

Uptake of cadmium by plants is greatest from soils high in cadmium
but is decreased by raising the pH by liming. Uptake varies with the soil
properties, especially those that control adsorption, and also with the
plant species and cultivar, and the source of cadmium. It has been
shown, for example, that uptake is less from soil to which sewage sludge
has been added than from soil to which an equal amount of cadmium
has been added as a salt.

Much of the cadmium that enters the plant is retained in the roots.
For example, in 20 out of 23 species grown in solution culture more than
half of the cadmium taken up was retained in the roots, the exceptions

being kale, lettuce and watercress. Oats, wheat and rice grown in soil retained 40–70% of the cadmium in their roots (Figure 10.2). Cadmium translocated to the shoots is present in greater concentration in the leaves than in fruits or seeds.

Lead

Lead was one of the first metals used by man. Because it is malleable it was used by the ancient Romans for making water pipes and other objects. The metal is used at the present day as a roofing material, in batteries and cable sheathing, and as lead shot. The metal has a very low solubility in acid and alkaline conditions, and is resistant to corrosion; hence these sources release very little lead into biological cycles. More important are lead compounds used as pigments in house paints and in the manufacture of plastics, glasses and glazes, although their solubility is low. Of greatest concern is the presence of lead in petrol, which is emitted from car exhausts and is the most widespread source of the contamination of soils and plants. Contamination also occurs near to, and downstream from, old lead mining areas, and by deposition from the atmosphere of particulate material released from smelters.

The toxicity of lead to humans has long been recognized. With the wider use of alternative materials, exposure to lead from water pipes, house paints, children's toys and food containers is becoming less; the use of unleaded petrol is also becoming more common. The lead is added to petrol as lead tetraethyl and tetramethyl and is emitted from car exhausts as $PbBrCl$ in particulate form. Most is deposited within 50 m of highways but small particles can be transported several kilometres. Either during transport through the air or after deposition, conversion to lead sulphate ($PbSO_4$) takes place. The greatest hazard to humans from this source of lead occurs through inhalation of the particles and consumption of leafy vegetables grown in urban gardens on which the particles have been deposited.

In acid solution lead is present as Pb^{2+} and with increasing pH it hydrolyses to form $Pb(OH)^+$ and $Pb(OH)_2^0$. It is strongly adsorbed by soils, the solution in uncontaminated soils having a concentration of about 2×10^{-6} M. Analysis of contaminated soils shows that lead is concentrated near the soil surface, very little moving down the soil profile. The low mobility is due to adsorption onto the surfaces of iron and manganese oxides and clay aluminosilicates. Lead also reacts with organic matter to form a complex of low solubility. Lead may be precipitated in near neutral and alkaline conditions as $Pb(OH)_2$ and $PbCO_3$, and also as lead sulphate.

The concentration of lead in soil solutions is low and uptake by plants from soils even near highways is also low. Additions of lead salts to soils or nutrient solutions produce only small increases in uptake, and most of this is retained in the roots. Lead is excluded from cereal grains, fruits and edible roots. The lead deposited on leafy vegetables in urban gardens is a greater hazard because only about one half is removed by washing.

Zinc

In the Earth's crust and soils zinc is the most abundant heavy metal that we are considering. Its greatest use is as a metal coating and in alloys. In moist air zinc metal is oxidized and forms a tough film of basic zinc carbonate which protects it from further chemical change. Iron coated with zinc is known as galvanized iron; the zinc protects the iron from rusting. The best known alloy of zinc is brass, which is the alloy with copper. Zinc is also used in batteries, and zinc oxide is used as a pigment in paints and as a filler in rubber, including tyres.

Zinc differs from cadmium and lead, two metals already discussed, in being an essential micronutrient for plants and animals, including humans. The supply of zinc from soils is inadequate for crops in some countries, especially in parts of Australia and the USA where increasing amounts of fertilizers enriched with zinc are applied to arable and horticultural crops.

At high concentrations in soil zinc can be phytotoxic, reducing the growth of plants or killing them at tissue concentrations that would not be toxic to animals. The principal hazard with zinc is therefore with plants, and problems are associated with the mining and smelting of zinc ores, the application of sewage sludge, which usually contains more zinc than any other metal known to harm plants, and sites beneath galvanized wire netting and electric cables. Where zinc toxicity occurs, liming the soil to increase the pH lowers the solution concentration, thus providing a remedy.

In acid solutions zinc is present as the divalent ion Zn^{2+}. In the pH range commonly found in soils (5–8) other ions are also formed including $Zn(OH)^+$, $Zn(OH)_2^0$, and $ZnHPO_4^0$, the extent of formation depending on the pH and concentration of HPO_4^{2-} in solution. When solutions of zinc salts are added to soil, Zn^{2+} exchanges with Ca^{2+} and other adsorbed ions. Much of the zinc is, however, strongly adsorbed, probably as $Zn(OH)^+$ onto iron and manganese oxides and clay aluminosilicates, adsorption increasing with pH. Zinc also forms complexes with soil organic matter, but these are less stable than those formed by cadmium, and much less stable than those formed by lead.

The ion species taken up by plant roots is not known with certainty but Zn^{2+} is generally assumed to be dominant. High amounts of zinc in soil may lead to high uptake, and on contaminated sites the concentration can exceed 500 µg Zn g^{-1} plant dry matter, a level considered to be generally toxic. Plant species differ in the extent to which they translocate zinc from roots to shoots; concentrations in fruits and grains are generally much lower than in leaves. Plants that retain zinc in their roots have been found to be more tolerant of high concentrations in the soil than those which transfer a high proportion to the shoots.

Fluorine

This is included among the elements discussed here to illustrate the point that toxicity is not limited to heavy metals. It is not the only non-metal that can cause toxicity. Boron, also a non-metal, can be toxic to plants when large amounts of fly ash (the dust produced in coal-fired power stations) is spread on soils, and when excessive amounts of a boron compound are applied to prevent a deficiency in plants.

Fluorine is not an essential element for plants but its beneficial effects on the incidence of dental caries is well established. High concentrations of fluorine or its compounds are, however, toxic to plants and animals.

Various industries, especially aluminium smelters, cement and brick kilns, and phosphate fertilizer factories emit fluorine compounds to the atmosphere in both particulate and gaseous forms. The particulate emissions are deposited close to the source and although they do not damage plants they may harm animals grazing affected pastures. Gaseous fluorides, principally hydrogen fluoride and silicon tetrafluoride, can harm plants and grazing animals, even when present in very low concentrations. Fluorine compounds are also emitted from some volcanoes, giving high concentrations in drinking water.

Fluoride ions (F^-) form complexes with aluminium in solution and on the surfaces of clay minerals and hydrated aluminium oxide. It is strongly held by the mineral surfaces so very little fluoride leaches through the soil in drainage water. Fluoride also reacts with calcium to form calcium fluoride, and with calcium and phosphate to form calcium fluorapatite, both having low solubility. Because of adsorption and precipitation the concentration of fluoride in soil solutions is low. Uptake by plants from soil is not significantly increased either by deposition from the atmosphere or by the presence of the small amounts present in phosphatic fertilizers.

10.6 Treatment of contaminated land

Contamination with sewage and at landfill sites will be avoided by following present regulations. Contamination by heavy metals that has already occurred can be dealt with in various ways:

(i)   Liming the soil to raise pH to 6.5 reduces the availability of most metals to plants for reasons given earlier.

(ii)   Addition of organic matter, which has the same effect as liming.

(iii)   Improved drainage can also be remedial because the most oxidized forms of iron (Fe^{3+}) and manganese (Mn^{4+}) oxides adsorb the metals.

(iv)   Tolerant plant species should be grown, if necessary using genotypes transferred from metalliferous sites.

(v)   Where contamination is extreme new topsoil should be added.

10.7 Radionuclides

Radionuclides are unstable isotopes which undergo radioactive decay. Some occur naturally in air, rocks, soils and plants at concentrations that give measurable amounts of radiation and some are produced artificially, as in nuclear weapon testing.

Becquerel discovered radioactive decay in 1896 when he found that a uranium salt blackened a photographic plate even in complete darkness. The work of Marie and Pierre Curie, Rutherford and Villard established three kinds of radiation called α (alpha), β (beta) and γ (gamma).

α-rays have the least penetrating power, being stopped by a sheet of paper or several centimetres of air. Each α particle consists of two protons and two neutrons and has a velocity about one tenth that of light.

β-rays are more penetrating than α-rays and pass through a sheet of aluminium. They can be either positively or negatively charged. A negatively charged β particle is an electron; a positively charged β particle is a positron. Each has a velocity approaching that of light.

γ-rays are the most penetrating, requiring several centimetres of lead to absorb them. They are not deflected in a magnetic field nor do they consist of particles, but are part of the electromagnetic spectrum; they are similar to high energy X-rays but with a shorter wavelength.

All three kinds of radiation are a health hazard. They can cause leukaemia, eye cataracts, various forms of cancer and, from a high dose, death. The hazard is greatest if the radionuclide enters the body through the intake of contaminated food, water or air. The effect of the radio-nuclide depends on (i) the extent to which it is retained, (ii) the energy level of its emitted radiation, (iii) its half-life, and (iv) the sites in the body where it is retained. α-particles are the most damaging to biological cells but because these particles have poor penetrative power they cause damage mainly when present internally, for example in the lungs if small particles of α-emitters are inhaled.

Units and definitions

1. Radioactivity is now measured in becquerel, Bq, where 1 Bq = 1 radioactive disintegration per second; the old unit, the curie, is equal to 3.7×10^{10} Bq.

2. The unit for the dose of radiation absorbed by body tissues is the gray, Gy, which corresponds to the absorption of energy equivalent to one joule per kilogram of tissue; the old unit, rad, is equal to 0.01 Gy. Also used is the sievert, Sv, which takes account of the radiation received by tissues modified according to the quality of the radiation and its distribution in the body.

3. The half-life is the time taken for a radionuclide to decay to one half of its initial value. As the rate of decay is exponential, a radio-nuclide with a half-life of 1 year leaves $\frac{1}{2}$ active after 1 year, $\frac{1}{4}$ after 2 years, $\frac{1}{8}$ after 3 years, and so on.

10.8 Radionuclides in the environment

A radionuclide that exists naturally in the environment is carbon-14, which is produced in the atmosphere by the bombardment of nitrogen (^{14}N) by cosmic rays. As ^{14}C has a half-life of 5.7×10^3 years it is present in all living plants and animals and soils but has almost completely decayed from fossil fuels. It is commonly used to date buried objects containing carbon.

Potassium-40, another naturally occurring radionuclide with a much longer half-life of 1.3×10^9 years, was produced when the universe was formed. Its concentration is now about 0.01% of the total potassium in rocks and soils. High concentrations of ^{40}K occur in rocks with high concentrations of total potassium such as granite and in soils derived from granite. Potassium fertilizers are obtained from natural deposits of potassium salts and their use increases the ^{40}K concentration in soils and plants. The ^{40}K passes into the food chain through all common foods, especially leafy vegetables and milk.

The gas radon-222 (^{222}Rn) has attracted interest and some concern in recent years because of its accumulation in residential buildings. It is a product of the decay of radium-226 (^{226}Ra) which occurs in small concentrations in rocks and soils; its highest concentration in commonly occurring rocks is in granite. Radon-222 decays to radioactive daughters, which attach themselves to dust particles, and these can be inhaled and deposited in the lungs.

There is intense radiation near deposits rich in uranium that are exploited commercially, in monazite sands containing thorium, and rocks that have high concentrations of radium. High concentrations of uranium also occur in rocks high in phosphate which is transferred to phosphate fertilizers, though in very small concentrations, and increases the load of radioactivity in agricultural soils.

All these sources, together with radiation from space, provide the continuous background radiation received by all forms of life on Earth. It has been estimated that, on average, 79% of the radiation to which humans are exposed is from natural sources, 19% is from medical applications and the remaining 2% is from fallout from weapons testing, television sets and the nuclear power industry. Although natural sources are dominant, most of the concern over radiation from radionuclides started with the development of nuclear weapons, after which increasing amounts of radionuclides were deposited on the Earth's surface.

The above-ground testing of nuclear weapons began in 1945. World-wide concern about radioactive fallout led to the agreement of 1963 between the USA, (then) USSR and UK to stop above-ground testing, although such tests have since been carried out by other countries. After the start of the nuclear power industry several accidents or poor standards of waste disposal have caused radionuclides to escape into the environment. The most severe accident was that at the Chernobyl nuclear power station in the (former) USSR on 26 April 1986, which caused several deaths at the power station itself and an unknown number in the surrounding area. The explosion created a radioactive plume containing several radionuclides, which passed over Scandinavia,

Table 10.6. *Characteristics of some radionuclides that occur naturally or are produced in nuclear reactors*

Isotope	Half-life	Principal radiation	Main occurrence
^{14}C	5.7×10^3 years	β^-	Natural and nuclear reactors
^{40}K	1.3×10^9 years	β^-	Natural
^{222}Rn	3.8 days	α	Natural
^{89}Sr	52 days	β^-	Nuclear reactor
^{90}Sr	28 years	β^-	Nuclear reactor
^{131}I	8 days	β^-, γ	Nuclear reactor
^{134}Cs	2 years	β^-, γ	Nuclear reactor
^{137}Cs	30 years	β^-, γ	Nuclear reactor
^{239}Pu	2.4×10^4 years	α, X-rays	Nuclear reactor

Poland and other northern European countries including the UK. The radioactive fallout was greatest where the plume density and rainfall coincided. Health hazards were created initially by the fallout of ^{131}I, ^{134}Cs, and ^{137}Cs; because of the longer half-life of ^{137}Cs (30 years) its effect will remain for many years.

There is now a vast array of radioactive substances manufactured for the nuclear power industry, nuclear weapons and for industrial, chemical and medical use. The radionuclides they contain have fission (daughter) products which are themselves often radioactive. Some of the most common radionuclides are listed in Table 10.6.

10.9 Radionuclides in soil

The radionuclides in soil that have caused concern have come from the testing of nuclear weapons, accidents at nuclear power stations, and poor waste disposal and storage.

A radionuclide undergoes the same reactions in soil as the non-radioactive isotope. Its adsorption depends on: (i) the properties of the ion in solution, and specifically whether it is negatively or positively charged or uncharged, and whether it forms complex ions in solution; (ii) the amounts in soil of the oxides of iron, manganese and aluminium and of the different clay aluminosilicates; and (iii) the amount and nature of organic matter that is present. Reducing conditions and pH affect the adsorption properties of soil, and also affect the ionic form of some radionuclides in solution. For very small amounts of an ion, as is usual for radionuclides in soil, adsorption occurs more commonly than precipitation. The latter tends to increase at pH above 6 by the formation of low soluble hydroxides and carbonates. The greater the retention of a radionuclide by soil the less it is taken up by plants and the less it is washed into drainage water.

The reactions of three radionuclides of caesium, iodine and strontium, will now be discussed. Each is an environmental hazard.

Caesium

The radionuclides ^{134}Cs and ^{137}Cs are produced in nuclear reactors and can be present in their discharge liquid. Both were released in large amounts to the atmosphere from the accident to the Chernobyl reactor and were deposited over much of Europe. Because of their long half-lives (Table 10.6) the isotopes are persistent. For several years after the accident and 2000 km from its location the movement and sale of sheep in parts of northern and western Britain have been restricted because of contamination of soil and pastures by caesium.

Caesium ions (Cs^+) are adsorbed by soils. When they displace more hydrated ions, e.g. Na^+, Ca^{2+}, Mg^{2+}, from the interlayer spaces of clay minerals like smectites and illite, the layers collapse and trap the caesium ions in cavities; see also Section 4.5. The adsorption of small amounts of Cs^+ by these clay minerals and especially by illite is difficult to reverse. For this reason the radionuclides of caesium in the fallout from this accident remain concentrated in the top few centimetres of uncultivated soils. Nevertheless some is taken up from the soil, especially from peats, by plants (which also received caesium directly on their leaves) and passes into the food chain.

Iodine

The radionuclide ^{131}I escaped into the atmosphere as a result of accidents to nuclear reactors at Windscale (now Sellafield) on the Cumbrian coast in the UK in 1957 and at Chernobyl in 1986. The fallout contaminated herbage eaten by dairy cows and was transferred to milk. Because of its short half-life of 8 days it presents a health hazard only when it quickly enters the food chain, as through herbage eaten by dairy cows and by the ingestion of leafy vegetables. Contamination of soils with ^{131}I is less of a hazard than with ^{134}Cs and ^{137}Cs because most will decay during the course of a growing season.

In the soil solution iodine is present as I^-. The ions are retained by soils by reaction with organic matter, oxides of iron and aluminium and clay aluminosilicates. Retention by the oxides increases at low pH due to the presence of positive charges. Another isotope, ^{129}I, is present in waste from reprocessing plants, and as it has a half-life of 1.7×10^7 years its retention in soil can be a long-term hazard.

Strontium

Of the two radionuclides of strontium, ^{89}Sr has a half-life of 52 days and ^{90}Sr a half-life of 28 years. Both have been released into the atmosphere from the testing of nuclear weapons and from accidents at nuclear power stations. It has been a cause of great concern because it behaves like calcium in the food chain, the radionuclides passing into bones from milk and other foods.

In solution strontium is present as Sr^{2+} and undergoes cation exchange reactions in a similar way to Ca^{2+}. It is readily taken up by plant roots and translocated to leaves, fruit and seeds. Uptake can be reduced, especially in soils low in calcium, by application of gypsum or lime.

Other radionuclides

Storage, to allow time for decay, is required for several radionuclides, including those from the production of nuclear weapons, the reprocessing of spent fuel from nuclear power plants, and the waste from medical and scientific laboratories. Several of them have long half-lives and storage must therefore be safe over a period of thousands of years to a standard that meets public requirements.

Some of these radionuclides, for example those of cerium, plutonium and ruthenium, are strongly retained by soils, the reaction depending on their ionic form in solution and on the properties of the soil. Plutonium is strongly adsorbed by reaction with organic matter and iron, manganese and aluminium oxides. Cerium ions (Ce^{3+}) and ruthenium ions (Ru^{3+}) are present in acid solution and are strongly adsorbed because of their high positive charge; at a higher pH Ce^{3+} precipitates as the hydroxide or carbonate. Technetium forms weakly adsorbed ions, TeO_4^-. Radionuclides that are strongly adsorbed by soil are not mobile and little will pass into drainage water.

10.10 Summary

Chemical elements that accumulate in soils at concentrations hazardous to plants or animals create a problem because of their persistence. This chapter describes the effects of heavy metals and radionuclides.

Heavy metals (arsenic, cadmium, chromium, copper, lead, mercury, nickel and zinc are considered the greatest hazard) accumulate in land-fill sites and from mining and smelting and the application of sewage sludge. They are adsorbed by soils and are generally retained in plant roots so that toxicity to plants occurs only at high concentrations in the soil and especially at low pH. Legislation now defines the maximum concentrations that are permitted.

Radionuclides are of concern because of the radioactive fallout from the testing of nuclear weapons, leakages including those during the long-term storage of radioactive waste, and accidents at nuclear power stations, for example that at Chernobyl. Radionuclides of caesium and strontium have long half-lives and although they are adsorbed by soils (caesium is only weakly adsorbed by peats) they can pass into the food chain where they represent a hazard to grazing animals and humans.

11

Soils, the atmosphere, global warming and ozone depletion

11.1 Introduction
11.2 The atmosphere: physical properties
11.3 The atmosphere: chemical properties
11.4 Radiative heating
11.5 Radiatively active gases
11.6 Carbon dioxide
11.7 Methane
11.8 Nitrous oxide and nitric oxide
11.9 Other gases
11.10 Changes of the global climate
11.11 Effects of global warming on soils
11.12 Summary

11.1 Introduction

The mean temperature of the Earth's surface is 288 K (15 °C). It is kept at this temperature by the so-called greenhouse effect without which it would be about 34 degrees lower. There is concern, however, that higher concentrations of CO_2 and other gases in the atmosphere will lead to higher temperatures at the Earth's surface, that is, to global warming. There is also concern that the emission of certain gases will decrease the layer of ozone in the stratosphere, which will lead to more ultraviolet light reaching the Earth's surface, as discussed in Section 11.3.

An important distinction has to be made between a greenhouse effect, which keeps the Earth's surface warm and at a roughly constant temperature, and an enhanced greenhouse effect, which results in global warming. The greenhouse effect is benign, but global warming is expected to lead to rising sea levels, changes of annual and seasonal rainfall at a regional scale, and to effects on vegetation systems and agriculture which are as yet uncertain. Rising sea levels will require substantial capital investment in barriers to prevent flooding and might cause serious loss of life, and a change of weather patterns might result in food shortages.

Although emission of CO_2 from the burning of fossil fuels is the main cause of the problem, there is an appreciable flux of CO_2 from the oxidation of soil organic matter and the burning of forests. Soil is also an important source of the greenhouse gases methane, CH_4, and nitrous oxide, N_2O. Before discussing what is known about the emission of the greenhouse gases from soil, an outline will be given of the physical and chemical properties of the atmosphere in order to explain the greenhouse effect.

11.2 The atmosphere: physical properties

On a volume per volume basis, described by atmospheric chemists as mixing ratios, the main gases in the Earth's atmosphere are nitrogen (0.781) and oxygen (0.209). The remaining 1% is made up of argon and other inert gases, water vapour, carbon dioxide, ozone and other gases, some of which will be referred to later. Small concentrations are usually expressed as parts per million by volume (ppmv) or parts per billion by volume (ppbv).

The total mass of the atmosphere is about 5×10^{18} kg. The innermost 10–17 km is the *troposphere* (Figure 11.1) in which mixing of gases is rapid. Above the troposphere for 30–40 km is the *stratosphere*, and between the two is the *tropopause*. The mixing ratios of the main gases, namely nitrogen and oxygen, are roughly the same throughout the atmosphere but the density decreases almost exponentially from the Earth's surface. The temperature generally decreases up to the tropopause and then increases up to a height of about 50 km.

A concept used by atmospheric chemists is the *residence time* (atmospheric lifetime) of gases emitted into the atmosphere. This is calculated as the amount of the gas in the atmosphere divided by its rate of removal. At a concentration of 1.72 ppmv the amount of methane in the atmosphere is about 4.3×10^{12} kg; if the annual rate of removal is 4.3×10^{11} kg, the residence time is 10 years.

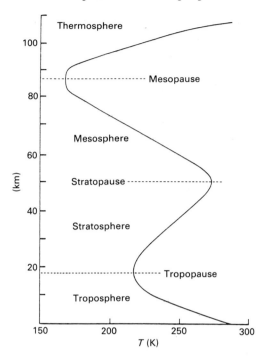

Figure 11.1 The structure of the atmosphere and its temperature profile. The temperature is given in degrees Kelvin (K) which is equal to degrees Celsius (°C) plus 273. (From Wayne, R.P., 1991. *Chemistry of Atmospheres.* Clarendon Press, Oxford; with permission.)

11.3 The atmosphere: chemical properties

Chemical processes in the atmosphere are often complex but only a few reactions need concern us here. Ozone, O_3, in the stratosphere is the starting point. It absorbs energy in the ultraviolet (UV) region of the spectrum and dissociates:

$$O_3 + h\nu \rightarrow O_2 + O, \qquad (11.1)$$

where $h\nu$ is the energy of one photon and the electrons in the products are in an excited state. Because of their high energy level, excited oxygen atoms initiate several important processes:

$$O + H_2O \rightarrow 2OH; \qquad (11.2)$$

$$O + CH_4 \rightarrow OH + CH_3; \qquad (11.3)$$

$$O + N_2O \rightarrow 2NO. \qquad (11.4)$$

The gas nitric oxide, NO, and the hydroxyl, OH, and methyl, CH_3, radicals are themselves reactive. During day time in the troposphere OH oxidizes several gases emitted from the Earth's surface. For example:

$$CO + OH \rightarrow H + CO_2; \tag{11.5}$$

$$SO_2 + 2OH \rightarrow H_2SO_4, \tag{11.6}$$

which are equations for the net reactions and do not show the intermediate products.

The radical CH_3 is oxidized to CO_2 and H_2O through a series of intermediates. Nitric oxide is first oxidized to NO_2 and then to HNO_3 (Equations 11.7 and 11.8). An important reaction of NO_2 in the troposphere is with the peroxyacetyl radical, $CH_3CO \cdot O_2$, which is formed during the incomplete oxidation of hydrocarbons. The product is peroxyacetyl nitrate, known as PAN. Ozone and PAN are the irritants in smog; they can cause health problems when their combined concentrations exceed 0.25–0.3 ppmv.

The comparatively stable gases, N_2O, chlorofluorocarbon compounds (CFCs) and to a lesser extent CH_4, diffuse into the stratosphere where their chemical reactions involve ozone. The reactions involving ozone and oxygen atoms in which the electrons are in an excited state are complex. For example, the nitric oxide formed from nitrous oxide (Equation 11.4) is oxidized further, leading to the destruction of ozone:

$$NO + O_3 \rightarrow NO_2 + O_2; \tag{11.7}$$

$$NO_2 + OH \rightarrow HNO_3. \tag{11.8}$$

CFCs, which are entirely man-made, also lead to the destruction of ozone, and on a bigger scale; for example, CFC-12 is partly decomposed by photons in the stratosphere:

$$CF_2Cl_2 + h\nu \rightarrow CF_2Cl + Cl. \tag{11.9}$$

One reaction is then

$$Cl + O_3 \rightarrow ClO + O_2; \tag{11.10}$$

$$ClO + O \rightarrow Cl + O_2; \tag{11.11}$$

net $\quad\quad O + O_3 \rightarrow 2O_2. \tag{11.12}$

Recent research has shown, however, that reactions involving chlorine in the Antarctic, which reduce the concentration of ozone and thereby create the so-called 'ozone hole', are more complex than these equations indicate. Most of the atmospheric ozone is found in the stratosphere. As mentioned below it absorbs nearly all the UV from the sun between wavelengths 0.2 and 0.3 μm. These wavelengths are harmful to all living cells and may cause skin cancers. For these reasons destruction of stratospheric ozone is viewed with great concern.

11.4 Radiative heating

Almost the whole of the energy input to the Earth is solar radiation. The radiation that reaches the Earth's atmosphere from the sun has wavelengths from about 0.2 to 4 μm, spanning the near ultraviolet (9%), visible (41%) and near infrared (50%). These wavelengths are determined by the temperature of the solar surface (about 6000 K).

Nearly all the ultraviolet (UV) in the solar radiation is absorbed by atmospheric gases. Oxygen and nitrogen absorb the solar UV with wavelengths up to 0.2 μm, and ozone absorbs UV between 0.2 and 0.3 μm. A small part of the visible radiation is scattered (which gives colour to the sky) and about 30% is reflected back to space by clouds and the Earth's surface. The rest heats rocks, soils and bodies of water. The air above the surface is then heated by conduction, convection and radiation; some of the absorbed solar radiation becomes latent heat of vaporization as water evaporates, but is later released to the atmosphere when condensation and precipitation occur.

Because the surface of the Earth is much cooler than that of the sun the radiation emitted by the Earth is of longer wavelengths (5–100 μm), referred to as the infrared. Whereas the atmosphere is relatively transparent to solar radiation, atmospheric gases absorb the longer wavelengths emitted from the Earth's surface. Several gases absorb in the infrared. The most important is water vapour, which is active, to a varying extent, throughout the infrared. Another is carbon dioxide, which absorbs between 12 and 17 μm. These gases emit the absorbed radiation at the same wavelength, and some of this radiation is emitted back to the Earth's surface. It is this long-wave radiation from the atmosphere which keeps the Earth's surface warmer than it otherwise would be.

The gases that absorb in the infrared are said to act like the roof of a greenhouse, hence the term 'greenhouse effect', although the comparison is false because the air inside a greenhouse is warmer than outside mainly because convective loss of heat is prevented. An example of the greenhouse effect is the absorption of infrared radiation by a blanket of clouds, which can prevent the formation of frost at night during the winter.

11.5 Radiatively active gases

Clouds, water vapour, CO_2, CH_4, N_2O, O_3 and the chlorofluorocarbons (CFCs) are the main components of the atmosphere that reduce the loss of outgoing infrared radiation. They are known as radiatively active gases ('greenhouse gases') and are listed in Table 11.1 together with their concentrations. Water vapour is not included because its concentration is very variable, nor is ozone because of lack of information on its trend of concentration with time.

The concentration of each of the gases is determined by the flux to and from the atmosphere and by chemical reactions in the atmosphere, many of which are complex. For example, methane is emitted from anaerobic soils and is absorbed by aerobic soils; its main sink, however, is the troposphere where it reacts with hydroxyl radicals. These radicals also react with other gases, including carbon monoxide, CO, and NO_x (NO_x = nitric oxide, NO, plus nitrogen dioxide, NO_2). Increasing concentrations of CO and NO_x, which are emitted by motor vehicles, are believed to lower the concentration of hydroxyl radicals and so decrease the rate of destruction of methane, leading to higher concentrations in the troposphere.

In addition to their radiative activity, some of the gases affect the concentration of stratospheric ozone. CFCs and N_2O are both radiatively active and both lead to the destruction of O_3 in the stratosphere (see Equations 11.1–11.4, and 11.7–11.12), as has been reported over the Antarctic, Arctic and elsewhere in recent years.

Also present in the atmosphere are aerosols, which are particles of diameter less than 3 μm that remain suspended in the air. They consist of man-made salts, e.g. ammonium sulphate formed largely from SO_2 from the burning of fossil fuels and NH_3 from organic manures, salts from sea-spray, particles from industrial emissions, soil particles result-

Table 11.1. *Summary of key greenhouse gases influenced by human activities*

	CO_2	CH_4	CFC-11	CFC-12	N_2O
Pre-industrial atmospheric concentration (1750–1800)	280 ppmv	0.8 ppmv	0	0	288 ppbv
Atmospheric concentration (1990)[a]	353 ppmv	1.72 ppmv	280 pptv	484 pptv	310 ppbv
Current rate of annual atmospheric accumulation	1.8 ppmv (0.5%)	0.015 ppmv (0.9%)	9.5 pptv (4%)	17 pptv (4%)	0.8 ppbv (0.25%)
Atmospheric lifetime[b] (years)	(50–200)	10	65	130	150

[a]The 1990 concentrations have been estimated based upon an extrapolation of measurements reported for earlier years, assuming that the recent trends remained approximately constant.
[b]For each gas in the table, except CO_2, the 'lifetime' is defined here as the ratio of the atmospheric content to the total rate of removal. This time scale also characterizes the rate of adjustment of the atmospheric concentrations if the emission rates are changed abruptly. The 'lifetime' of CO_2 given in the table is a rough indication of the time it would take for the CO_2 concentration to adjust to changes in the emissions.
Note: ppmv, parts per million (10^6) by volume; ppbv, parts per billion (10^9) by volume; pptv, parts per trillion (10^{12}) by volume.
Source: From Houghton, J.T., Jenkins, G.J. and Ephraums, J.J. (eds) 1990. *Climate Change, Intergovernmental Panel on Climate Change.* Cambridge University Press, with permission.

ing from wind erosion, and the products of volcanic eruptions. Their influence on global warming is not properly understood. Their main effect is probably to reflect solar radiation, reducing energy input and thereby cooling the planet. They might also affect the properties of clouds in ways that are being investigated.

Global warming is attributed to increased atmospheric concentrations of CO_2, CH_4, CFCs, and N_2O, and tropospheric O_3. Their relative contributions to global warming, calculated from their increased concentrations in the 1980s and ability to absorb infrared radiation, are shown in Figure 11.2. Soil is a source of CO_2, CH_4 and N_2O, as will be discussed next.

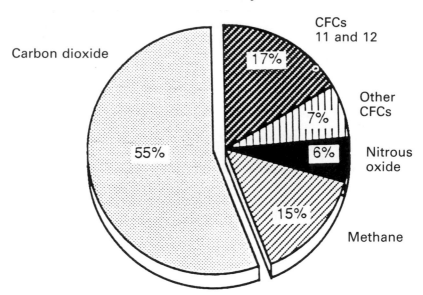

Figure 11.2 The contribution of greenhouse gases to global warming in the 1980s. (From Houghton J.T., Jenkins, G.J. and Ephraums, J.J. (eds.) 1990. *Climate Change, Intergovernmental Panel on Climate Change*, Cambridge University Press; with permission.)

11.6 Carbon dioxide

The pre-industrial concentration of CO_2 in the atmosphere, which is known from the concentration in air trapped in ice, was 280 ± 10 ppmv. When accurate monitoring started in 1957 it had risen to 315 ppmv. Between 1976 and 1982 the mean annual increase at three stations in the Pacific and at the South Pole was 1.5 ± 0.2 ppmv, that is, $0.4 - 0.5\%$ per year. By the late 1980s the concentration had reached about 350 ppmv. The data from the Mauna Loa Observatory in Hawaii (Figure 11.3) show this trend. They also show a regular seasonal variation, which is attributed largely to the photosynthesis of terrestrial vegetation in the northern hemisphere (it has a greater land area than the southern hemisphere) and to the decomposition of leaf litter and soil organic matter.

It has been calculated that the increase of 9 ppmv between 1976 and 1982 represented 54% of the known emissions of CO_2 in that period from fossil fuel combustion; cement manufacture also contributes to the emissions. The airborne fraction becomes lower than 50% when the emissions of CO_2 from soils and forest burning are included, but as discussed below these emissions are rather poorly quantified. An accu-

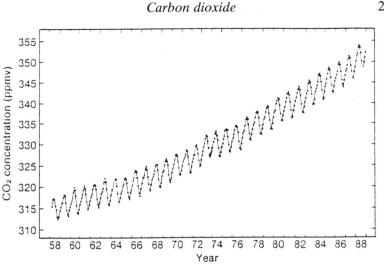

Figure 11.3 Monthly average CO_2 concentration in parts per million of dry air, observed continuously at Mauna Loa, Hawaii. (From Houghton *et al.*, 1990 (see Figure 11.2); with permission.)

Table 11.2. *The global carbon budget*

	Carbon (Gt a^{-1})
Emissions from combustion of fossil fuels	5.4 ± 0.5
Emissions from deforestation and land use	1.6 ± 0.5
Accumulation in the atmosphere	3.4 ± 0.2
Uptake by the oceans	2.0 ± 0.8
Net imbalance	1.6 ± 1.4

Source: From Houghton *et al.*, 1990 (see Table 11.1), with permission.

rate value for the airborne fraction, and especially whether it changes with rising CO_2 concentrations, is important for prediction of the extent of global warming. It is believed that the oceans are the sink for the CO_2 which does not remain airborne, but the quantity taken up by the oceans is still a matter for speculation.

The annual emissions from fossil fuel combustion are documented and are about 5.4×10^{15} g $CO_2 - C$. An estimate for the global carbon budget is given in Table 11.2. It shows a significant imbalance, which cannot yet be explained. One possibility is that photosynthesis is being increased by the raised CO_2 concentrations; another is that plant growth is being stimulated by the increased amount of nitrogen and possibly sulphur compounds being deposited from the atmosphere, and by the use of organic manures and fertilizers.

The effects of deforestation and oxidation of soil organic matter will now be discussed.

Soil carbon and CO_2 fluxes

The total amount of carbon in the organic matter of soils throughout the world is about 1500×10^{15} g, which is about the same as the combined total of carbon in terrestrial vegetation and in atmospheric CO_2 (Figure 11.4). Additionally, soils often contain charcoal, which is comparatively inert, and some contain large amounts of $CaCO_3$ and $MgCO_3$.

Oxidation of soil organic matter provdes energy for the soil organisms; CO_2, the product of respiration, is released into the atmosphere. If all the carbon in soil organic matter were released into the atmosphere as CO_2, the concentration in the atmosphere would double (assuming an airborne fraction of 0.5).

In a stable ecosystem with unchanging environmental conditions the organic carbon content of soil reaches a steady state. The rates of addition and loss of carbon become equal and the content of organic carbon is at equilibrium (Section 3.4). The equilibrium content becomes less when the soil is cultivated and will change if the environmental conditions change. A change in the soil organic carbon content will affect the CO_2 concentration in the atmosphere.

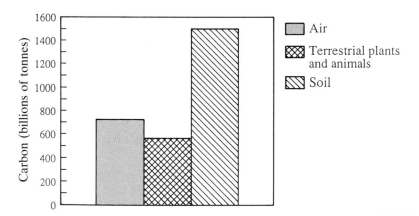

Figure 11.4 Amounts of carbon in soil organic matter, terrestrial vegetation and animals, and the atmosphere. (From Bolin, B. *et al.*, 1986. *The Greenhouse Effect, Climate Change and Ecosystems*, SCOPE 29, Wiley, Chichester; and other sources.)

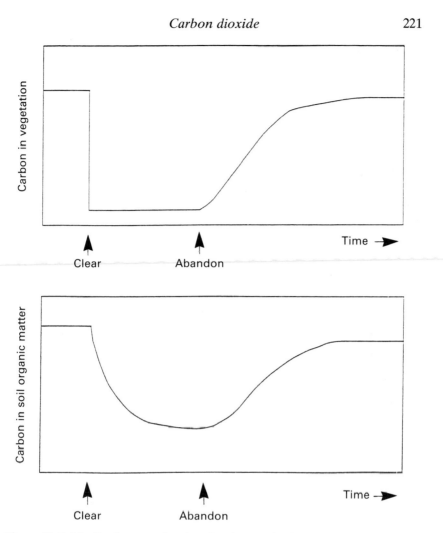

Figure 11.5 Idealized curves showing the changes in the carbon content of soil and vegetation after the clearance of land for agriculture, and after its abandonment and reversion to forest.

An example of accumulation of organic carbon content of soils is shown in Figure 3.1*b*. Rates of change will be seen to be rapid, so that carbon content can halve or double over a period of a few decades. The changes that occur when forest is cleared for agriculture are shown schematically in Figure 11.5. Carbon in the vegetation decreases immediately and that in soil more slowly; both increase again when the forest is regenerated.

The global release of CO_2 to the atmosphere as a result of conversion

Table 11.3. *Carbon contents of two global forest ecosystems, undisturbed, during cultivation and after regrowth*

	Tropical moist forest	Temperate deciduous forest
Area in 1700 (Mha)	1352	612
Carbon in vegetation of undisturbed forest (t ha^{-1})	200	135
Carbon in soils of undisturbed forest (t ha^{-1})	117	134
Carbon in crops during cultivation (t ha^{-1})	5	5
Minimum carbon in soil during cultivation (t ha^{-1})	58	67
Carbon in vegetation of secondary forest (t ha^{-1})	150	100
Carbon in soils of secondary forest (t ha^{-1})	88	120

Source: From Houghton, R.A. *et al.*, 1983. *Ecological Monographs* **53**, 235–262.

of natural ecosystems to agriculture is not known with any certainty. Estimates of the carbon content in the vegetation and soils of undisturbed tropical and temperate forests and during cultivation are given in Table 11.3. Using the same information for all ecosystems and the estimated extent of conversion to agriculture, it has been calculated that up to 1960 the net emission of CO_2 from vegetation and soils had exceeded that from the combustion of fossil fuels. Currently the annual contribution from deforestation and changing land use is believed to be 1.6×10^{15} g CO_2–C, that is, about 23% of the total emission of 7×10^{15} g CO_2–C to the atmosphere. The net emission from soils is probably less than from vegetation and is likely to be in the range from 0.2 to 0.5×10^{15} g CO_2–C annually.

11.7 Methane

The concentration of methane, CH_4, in air trapped in ice more than 300 years ago is about 0.8 ppmv. The concentration in the atmosphere in 1990 was 1.72 ppmv and is increasing at a rate of 0.9% per year. Since 1980 it has been the third largest contributor to global warming (Figure 11.2). Not only does it absorb infrared radiation, it is of concern because it also has adverse chemical effects in the atmosphere by reacting with OH radicals, and it has a long residence time of about 10 years. About 43% of the emission of CH_4 into the atmosphere is from natural wetlands and paddy fields (Table 11.4).

Part of the emitted CH_4 is oxidized in the troposphere to H_2, CH_3 and

Table 11.4. *Estimated sources and sinks of methane*

	Annual release (Mt CH_4)	Range (Mt CH_4)
Source		
Natural wetlands (bogs, swamps, tundra, etc)	115	100–200
Rice paddies	110	25–170
Enteric fermentation (animals)	80	65–100
Gas drilling, venting, transmission	45	25–50
Biomass burning	40	20–80
Termites	40	10–100
Landfills	40	20–70
Coal mining	35	19–50
Oceans	10	5–20
Freshwaters	5	1–25
CH_4 hydrate destabilization[a]	5	0–100
Sink		
Removal by soils	30	15–45
Reaction with OH in the atmosphere	500	400–600
Atmospheric increase	44	40–48

[a]Release of methane from the permafrost in northern latitudes.
Source: From Houghton *et al.*, 1990 (see Table 11.1), with permission.

CO, and part escapes to the stratosphere where it destroys ozone (Section 11.3), and where it is a source of water which may increase stratospheric clouds. Its reactions in the troposphere probably reduce OH concentrations, which may increase its persistence. Its total oxidation is to CO_2 and H_2O, but the reactions that produce these end products are complex and depend on the atmospheric concentration of nitrogen oxides (NO_x).

The flux of methane from soil

Methane is produced by biological processes under wet conditions where oxygen is excluded (anaerobic conditions) and for this reason used to be known as marsh gas. Methane-producing (methanogenic) bacteria can use several substrates formed by the decomposition of plant material; acetate, and a mixture of H_2 and CO_2, appear to be the most important in soils. It is only produced in a strongly reducing medium ($E_h < 100$ mV) and is therefore normally inhibited by SO_4^{2-}, as is present in marine marshes, which maintains a higher E_h. As methane can be oxidized in water that contains dissolved oxygen, the amount that reaches the surface depends on the depth of water. Its rate of production generally increases with temperature.

The methanogenic bacteria have an optimum pH of about 7 so that CH_4 emissions are low from acid swamps. Soil that is aerobic can be a sink for atmospheric CH_4, though it is probably not an important one.

The few field measurements of CH_4 emissions from soils and swamps have been made by a static box method. In paddy fields the box is floated on water over rice plants, and the air inside is sampled after short time intervals (30–60 min) for CH_4 analysis. Measurements on a rice paddy in Italy showed that CH_4 was emitted from the time of flooding, with highest rates at the time of tillering and flowering. Emissions were higher in the presence of plants than in their absence, suggesting that carbon compounds released from the plant roots were substrate for the methanogenic bacteria. There is evidence that some of the CH_4 is conducted to the atmosphere through the aerenchyma of plant roots and stems.

The box method of measurement has shown high temporal and spatial variability in CH_4 emissions. The problem of spatial variability can be lessened by determining the average concentration over distances of several metres using laser beams tuned to the infrared absorption of CH_4. The method is sensitive but because instrumentation is expensive it has not been widely used.

Estimates of global emissions of CH_4 from paddy fields and swamps are constantly being revised as new measurements are reported, and those given in Table 11.4 may well be changed. Apart from the paucity of measurements, especially from swamps, as yet there is no model that incorporates the environmental conditions controlling the production and emission of methane. The areas under swamp, and the different types of swamp are also not well known.

As mentioned above, the atmospheric concentration of CH_4 has been increasing at a rate of about 0.9% per year. This increase is generally attributed to the increase in area of rice grown on flooded soils (20% increase between 1960 and 1980), to leakage from natural gas production, to increased numbers of domestic cattle and sheep, and to increased burning of vegetation.

11.8 Nitrous oxide and nitric oxide

The present (1990) concentration of nitrous oxide, N_2O, in the atmosphere is 310 ppbv and is increasing by about 0.25% per year. It has a

Table 11.5. *Sources and sinks of nitrous oxide*

	Range (Mt N a^{-1})
Sources	
Oceans	1.4–2.6
Soils (tropical forests)	2.2–3.7
Soils (temperate forests)	0.7–1.5
Combustion	0.1–0.3
Biomass burning	0.02–0.2
Fertilizers (including ground-water)	0.01–2.2
Total	4.4–10.5
Sinks	
Removal by soils	?
Reactions in the stratosphere	7–13
Atmospheric increase	3–4.5

Source: From Houghton *et al.*, 1990 (see Table 11.1); with permission.

long residence time in the atmosphere of about 150 years. It absorbs infrared radiation but because of its low concentration it has only a small effect on global warming. Because of its long residence time it has, however, more serious effects in the stratosphere where it reacts with excited atomic oxygen to give nitric oxide, NO, which destroys ozone (Equations 11.4 and 11.7). Estimates of the sources and sinks of N_2O are given in Table 11.5.

The flux of nitrogen oxides from soil

Nitrous oxide is produced mainly by the biological reduction of nitrate, a process known as denitrification. The process, which is described in Section 5.7, occurs under anaerobic conditions. It takes place most rapidly when the soil is warm and contains readily decomposed organic matter. Microorganisms use NO_3^- instead of O_2 as an electron acceptor, producing NO, N_2O and N_2.

Nitrous oxide and nitric oxide are also released during nitrification. The process of nitrification is the oxidation of NH_4^+, described in Section 5.6, and can be written as follows:

$$NH_4^+ \rightarrow (NH_2OH) \rightarrow (NOH) \rightarrow (NO_2^-) \rightarrow NO_3^-.$$

The intermediates in brackets are unstable and either chemically or through enzyme activity yield products that include N_2O and NO. Rates of production of N_2O by this process are lower than by biological denitrification, but as most soils of the world are well aerated for most of the time, the process is believed to make a significant contribution to the global flux of N_2O and NO. The global flux of NO is not yet known.

Measurements of N_2O fluxes have been made by a variety of methods. They include the use of a closed cover placed on top of the soil, the air within it being sampled after short intervals of time; alternatively air can be drawn through a cover, the N_2O being trapped on a molecular sieve. In a modification of the closed-cover method, soil cores taken from the field are placed in jars which are made air-tight and the air inside is sampled for N_2O after short time intervals. The concentration of N_2O is measured by gas chromatography.

As with CH_4 there is great spatial and temporal variability in the emission of N_2O from soils. In one study 350 measurements were needed on a 108 m^2 plot for the sample mean to be within $\pm$ 10% of the true mean. The tuned laser beam method referred to above for CH_4 determination can also be used for N_2O to lessen the problem of spatial variability.

In order to extrapolate from sites where measurements have been made a quantitative description is needed of the factors that control N_2O emissions. The factors are aeration (or water content) of the soil, temperature, nitrate concentration, readily decomposable organic matter and pH. Equations have been fitted to existing data, but none have been widely tested under field conditions.

Soil temperature and water content affect the production of N_2O after application of ammonium nitrate as fertilizer. This was shown in a field experiment on grassland in southern England, in which four applications of fertilizer were made during spring and summer, each supplying 62.5 kg ha^{-1} of N. Production was highest in July when the soil was warm and wet; it was less in April–May and August–September when the soil was cooler and drier respectively. The highest rate was 212 g N_2O-N ha^{-1} d^{-1}. Between periods of peak production the soil acted as a weak sink, absorbing up to 10 g N_2O-N ha^{-1} d^{-1}.

Reported losses of nitrogen as N_2O vary from less than 1 kg ha^{-1} a^{-1} to over 100 kg ha^{-1} a^{-1}. The highest values are from irrigated and fertilized soils high in organic matter. From uncropped, unfertilized land in temperate regions the annual loss has been put at 0.2 kg ha^{-1}, and at 1–2% of fertilizer nitrogen. As the annual world use of fertilizer nitrogen is about 60 million tonnes, a loss of 1.5% would yield about 1 Mt of N as N_2O annually. The few measurements reported from unfertilized tropical forests indicate these to be a major source, possibly leading to an annual production of between 6 and 8 Mt of N as N_2O from this ecosystem alone. If the production rate from tropical forests is confirmed, the figures for global emissions will need to be revised upwards.

Nitrous oxide is a sparingly soluble gas. Small amounts can be removed from soil in drainage water and might later pass into the atmosphere. In one experiment 0.25 kg N_2O-N ha^{-1} was lost by this means in a 5-month winter period when 0.15–0.9 kg ha^{-1} was emitted from the soil surface into the atmosphere.

Nitric oxide is also produced in soils. As mentioned above it is a byproduct in nitrification and is also produced by the chemical decomposition of nitrite under acid conditions. From the few field measurements so far reported it seems that emissions of NO might exceed those for N_2O. In the atmosphere it is oxidized to NO_2 (Equation 11.7).

11.9 Other gases

Carbon monoxide is present in the troposphere at a low concentration of 110 ppbv, which might be increasing. Its residence time is about 0.3 years. Although not significantly absorbing in the infrared, CO scavenges atmospheric OH which can lead to increased concentrations of tropospheric O_3, CH_4 and other hydrocarbons. Its main source is the incomplete oxidation of fossil fuels, but some is produced by plants during photorespiration. Some processes in soils, possibly auto-oxidation of organic compounds, also yield CO. Many soil organisms oxidize CO to CO_2 and some may utilize the carbon for growth, making soil a sink for atmospheric CO.

Hydrogen is present in the atmosphere at a concentration of 550 ppbv; its residence time is about 2 years. It has a harmful effect because it reduces the concentration of OH in the troposphere. Soil is not an important source and is more usually a net sink. It is, however, produced at anaerobic sites in soils, although it is not necessarily emitted into the atmosphere because it can be consumed by methanogenic bacteria and by aerobic bacteria where oxygen is present. Emission in oxic environments is probably limited to some strains of *Rhizobium* during fixation of nitrogen in legume roots. Emission seems not to occur from other nitrogen fixing organisms.

Soils are a source or sink for other gases which affect atmospheric chemistry. Ammonia is emitted from animal excreta, manures, soil organic matter, and from fertilizers that contain urea or contain ammonium salts; emissions are greatest from soils that have a pH above 7–7.5 (see Section 8.5). It forms aerosols of ammonium sulphate and other salts, which can increase the albedo of the atmosphere. Its residence time in the atmosphere is a few days only, but some might be oxidized in the troposphere to oxides of nitrogen.

Reduced forms of sulphur including hydrogen sulphide and dimethyl-sulphide are emitted from anaerobic soils. They can be oxidized in the atmosphere to sulphate, thus leading to the formation of aerosols of ammonium sulphate referred to above.

11.10 Changes of the global climate

Changes in the past

Since the end of the last Ice Age about 10 000 years ago the average global temperature has differed from the present by between 1 and 2 degrees. During the past 2 million years (the Pleistocene) the temperature has varied by 5–7 deg. Regionally, the fluctuation of temperature has been greater. In northern latitudes the temperature may have fluctuated by 10–15 deg during the Pleistocene, and at these northern latitudes 5000–6000 years ago the summer temperatures might have been 3–4 deg higher than at present.

There is uncertainty about the exact changes of past temperatures, globally and regionally. This is because they are based on inferences from the vegetation shown by pollen remains and tree rings, the extent of glaciers and from the composition of ocean sediments. Since 1900 when temperatures have been measured directly, although not with global coverage or consistent instrumentation, the global increase has been 0.45 ± 0.15 deg (Figure 11.6). This increase cannot yet be attributed unambiguously to increased concentrations of greenhouse gases.

The changes of temperature in the past are attributed to various causes including the distance of the Earth from the sun and a change in tilt of the Earth's axis. Chemical analysis of the air trapped in ice has shown that the CO_2 concentration of the atmosphere was low during the last Ice Age. This might suggest that the Ice Age was caused by the low CO_2 concentration, which would lessen the greenhouse effect. Equally possible, however, is that the Ice Age had another cause and that the

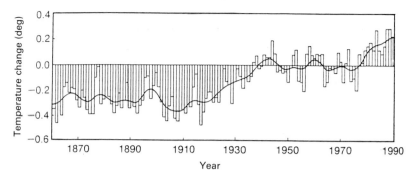

Figure 11.6 Global-mean combined land–air and sea surface temperatures, 1861–1989, relative to the average for 1951–80. (From Houghton *et al.*, 1990; see Figure 11.2; with permission.)

CO_2 concentration was low because more was removed from the atmosphere by its greater solubility in water at low temperatures, which would have accentuated but not have caused the Ice Age. The trigger for temperature change in the past remains largely conjectural.

Changes in the distribution of rainfall have also occurred. The Sahara Desert supported subtropical vegetation and animals from 12 000 to 4000 BP. Wetter conditions also prevailed from about 5000 to 6000 BP in the Arabian Peninsula and the southern parts of the Asiatic USSR. In the same period the rainfall was less in the eastern and central USA.

Predictions for the future

The Intergovernmental Panel on Climate Change (1990) suggested that the global mean temperature may increase during the next century by 0.3 deg per decade, with an uncertainty range of 0.2–0.5 deg per decade. This estimate assumes that the only change to present practices is a reduction in the emission of CFCs (the so-called Business-as-Usual scenario). The increase would be about 0.1 deg per decade if there were a big reduction in the emission of greenhouse gases. The effects on rainfall are uncertain, but one possibility is for regional changes similar to those that occurred 5000–6000 BP and referred to above.

11.11 Effects of global warming on soils

If emissions of radiatively active gases continue largely unchecked it is predicted that the 0.3 deg rise per decade will raise sea levels by about 6

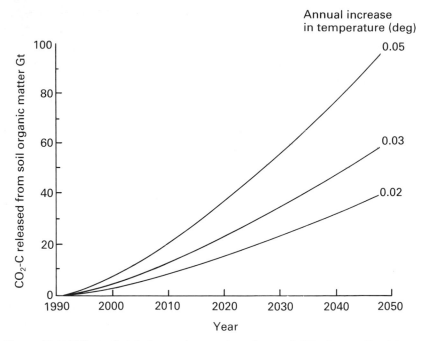

Figure 11.7 Effect of global warming on the release of CO_2 from soils. (From Jenkinson, D.S., Adams, D.E. and Wild, A. 1991. *Nature* **351**, 304.)

cm per decade, and the atmospheric CO_2 concentration will have doubled by the second half of the twenty-first century. The effects will be less if emissions of the gases are reduced. The predictions are not certain and can be expected to change as more understanding is gained of the global energy budget.

The effects on soil properties and processes cannot yet be predicted, but certain changes seem probable. These are:

1. A temperature rise will increase the rate of soil processes including mineral weathering, oxidation of organic matter and other biological processes, evaporative loss of water and diffusion of gases into the atmosphere.

In general, rates roughly double for each 10 deg rise of temperature so the effect will not be large. For example, it has been calculated that a global rise of 0.3 deg per decade, with no change of rainfall or increase of net primary production, will release an extra 61×10^{15} g CO_2-C from soil into the atmosphere by the year 2050 (Figure 11.7). This represents about 19% of the CO_2 that will be released by combustion of fossil fuels if their use remains unchanged.

2. Doubling of the CO_2 concentration from 350 to 700 ppmv is expected to increase the yield of C_3 crops (e.g. rice, soyabean, wheat) by between 10% and 50%, and C_4 crops (e.g. maize and sugarcane) by up to 10% (see Section 7.2 for description of C_3 and C_4 plants). Increased rates of growth of forests and other natural ecosystems might also occur. The supply of nitrogen often limits plant growth in natural ecosystems, and as it is now being circulated through the atmosphere in greater quantities, it might further increase net primary production. There might therefore be an increased addition of plant residues to the soil to counterbalance the increased rate of mineralization.

3. The regional patterns of rainfall and evaporation might change, some areas becoming wetter and some drier, but it is uncertain what these changes will be. Drier conditions will be ameliorated to some extent by the increased concentration of CO_2 which reduces water loss by transpiration, that is, it increases the efficiency of water use by plants. A change of rainfall will affect soil properties but it is too early to make sensible predictions.

4. A rise in temperature is expected to lead to a rise in sea level and so continue the trend from the past (Figure 11.8). On the Business-as-

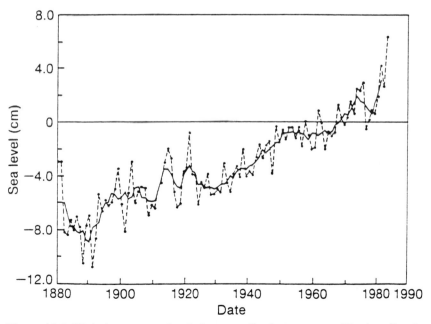

Figure 11.8 Global mean sea-level rise over the last century. The baseline is obtained by setting the average for the period 1951–70 to zero. (From Houghton *et al.*, 1990 (see Figure 11.2); with permission.)

Usual scenario the sea level is predicted to rise by about 6 cm per decade during the next century. Such a rise will affect all countries with a coastline, countries such as Bangladesh and Egypt being seriously affected because a sizeable fraction of their most productive land will be liable to be flooded. Low-lying land on the fringes of coasts and estuaries and some atolls will become submerged. Ground waters will rise near coasts in response to a rise in sea levels and where these ground waters are saline the soils will also become saline.

11.12 Summary

Global warming is attributed to increased concentrations of carbon dioxide, chlorofluorocarbon compounds, methane and nitrous oxide in the atmosphere. The destruction of ozone in the stratosphere is attributed to chlorofluorocarbon compounds, nitrous oxide and to a less extent methane. Soil is one of the sources of these gases, except for the chlorofluorocarbon compounds, and is the main source of nitrous oxide and methane. Each is produced in anaerobic conditions in soil, and nitrous oxide is also produced during nitrification.

Carbon held in the soil organic matter of the Earth is about twice the amount present in the atmosphere as carbon dioxide. About one third is oxidized to carbon dioxide within a few decades when land is brought into cultivation, and some is oxidized more slowly. Global warming will increase the rate of oxidation. It will also raise sea levels, which will submerge low-lying land near sea coasts. Land use will change because of the increased temperature and the probable change of rainfall distribution.

12

Soil erosion and conservation

12.1 Introduction

Soil erosion is the removal of part of the soil, or the whole soil, by the action of wind or water. It is a natural process that occurs without human intervention but it can be greatly increased by cultivation of the land. Indeed, there has been serious erosion as long as there have been farmers. In the northern mountains of ancient Mesopotamia soil erosion is associated with the early, and probably the first, farmers about 10 000 years ago. It is believed to have been caused by the felling of trees, cultivation of the cleared land and overgrazing by sheep and goats. The same causes led to erosion of soil from the highlands in Ethiopia, Greece and Italy. Erosion in northern Europe is believed to have begun with the clearing of woodlands about 5000 years ago. European settlers in North and South America and Australia were generally unaware of the potential hazards of clearing and cultivating land that was subject to drought and severe rainstorms. Water erosion, and wind erosion of the kind that occurred in the Great Plains of the USA in the 1930s, had devastating effects. The clear-felling of forests on steep slopes and in the humid tropics followed by cultivation is continuing to cause erosion which is at least as serious as any that has occurred in the past.

The results of erosion may not be entirely negative. For example, sediment derived from erosion in Ethiopia helped to sustain Egyptian agriculture for thousands of years. The same effects ensured the fertility of the Ganges–Jumna plain in India, although this is now threatened because excessive felling of trees in the Himalayas has led to excessive run-off of water and deposition of sediment which, in turn, have led to serious flooding in Bangladesh.

12.2 Natural erosion

Natural, or geological, erosion has several effects. It has occurred throughout the history of the Earth from the time that rocks were first exposed to the influence of the atmosphere, shaping the land surface and forming sedimentary rocks from weathered rock and soil.

Water erosion of hill slopes has widened valleys and produced colluvium on the foot slopes and alluvium in valley bottoms. (The amount of soil entering rivers is therefore less than the amount eroded.) Eroded material is carried by rivers out to sea where the coarser particles form deltas.

As a final example of natural erosion, wind-blown fine sand, known as loess, has been deposited in thick sheets in China, Europe and North America, forming some of the Earth's most fertile soils.

The rate at which natural erosion occurs depends on the vegetation cover of the land surface, its slope, the size of rock fragments and soil particles at the surface, and the climatic conditions. These factors are discussed in Sections 12.4 and 12.5 in relation to accelerated erosion, which is caused by man. Because of its variability an average value for natural erosion may have little relevance to a particular site in a particular year, but its order of magnitude is useful in providing the baseline from which to assess accelerated erosion, as discussed later.

12.3 The environmental problem

In many countries accelerated erosion is the most serious form of soil degradation. As for the extinction of plant and animal species, the loss of soil can be avoided but not reversed.

Erosion has, however, effects on the environment which are more extensive than damage at the site from which soil has been lost (Table

Table 12.1. *Effects of soil erosion*

1. Loss of soil to support the growth of crops, grassland, and forests
2. Silting-up of dams
3. Deposition of sediment loads causing rivers to change course
4. Variable seasonal flow of rivers, and flooding
5. Water pollution: erosion of 1 t of soil containing 0.2% N and 0.05% P will transfer 2 kg N and 0.5 kg P to rivers and lakes
6. Air pollution: fine soil particles in the air reduce solar radiation at the surface of the Earth and might affect chemical processes in the atmosphere

12.1). Soil particles that have been detached by water erosion may be transported in rivers. Their deposition as sediment can change the course of the rivers and cause flooding. If transported to the sea the deposit can damage corals and other marine life. Deposition in lakes, reservoirs and dams decreases their storage capacity and reduces their useful life. The sediment carries nutrients, which may cause eutrophication of rivers and lakes, and it may carry pesticides from agricultural land. These polluting effects increase the cost of water treatment for domestic use.

Loss of soil from hill slopes increases the surface run-off of water, which leads to flooding in the lower parts of the catchment during rain storms and to less flow in rivers during dry spells. Run-off water also carries water-soluble nutrients to rivers and lakes.

Wind-blown sand damages crops, can block waterways and can bury buildings. The fine particles that are carried into the atmosphere form a haze, which intercepts sunlight and can cause respiratory problems in humans; the particles may also act as nuclei for the formation of aerosols, which affect chemical processes in the atmosphere (Chapter 11).

Finally, accelerated erosion increases the rate of transfer of nutrients from land masses to the oceans. The effects of these nutrients on the productivity of marine life are not yet known.

Extent of erosion

Virtually all countries are affected to some extent by erosion. It is most severe after removal of the protective cover of vegetation by tree felling or cultivations or as a result of drought. Loss of the vegetative cover also increases the surface run-off of water; where water is a limiting factor for crop growth, soil and water conservation must be considered together.

Water erosion is severe in parts of the tropics, especially on steep slopes, because of heavy rainstorms. In the semi-arid and arid regions of the world, wind erosion occurs frequently, the soils themselves often being formed from wind-blown material. Erosion can, however, vary greatly from one year to another depending on the severity of rain storms or of drought.

Amounts of soil loss

Estimating the amount of soil erosion is more difficult than it seems. For example, the sediment load of rivers gives an underestimate because it does not include soil eroded by wind, nor the eroded soil that is deposited before it reaches the rivers. Estimating the reduction of soil depth is insensitive and can be used only when there has been substantial erosion. Observations in the field can be made of the occurrence of gullies and smaller channels (rills) formed by surface run-off of water, or of the frequency of dust storms and the effect of splash erosion. Such observations give estimates of the area of land that is affected but not of the amounts of soil that are lost.

Because of these difficulties, soil loss by water erosion is measured on experimental plots. Wind erosion can not be measured in a similar manner, but can be simulated by using wind tunnels. Using these methods, the factors that determine the amount of erosion can be investigated. These factors are described in Sections 12.4 and 12.5.

In determining the loss of soil by water erosion, the sediment from an experimental plot is dried and weighed; surface run-off is also collected and measured. Using this method at the International Institute for Tropical Agriculture, Ibadan, in southern Nigeria, the effects of slope and mulching on soil loss and run-off were measured (Figure 12.1). On experimental plots in Africa soil loss has been in the range 0 to 170 t ha^{-1} a^{-1}. Several factors contribute to the rate of erosion, as will be referred to in later sections.

Now to be discussed are water and wind, the main agents of soil erosion in humid regions and in arid and semi-arid regions, respectively. A more localized agent is frost heave on the slopes of mountains: the soil expands at right angles to the slope when it freezes, and stones and loose aggregates fall downslope when it thaws.

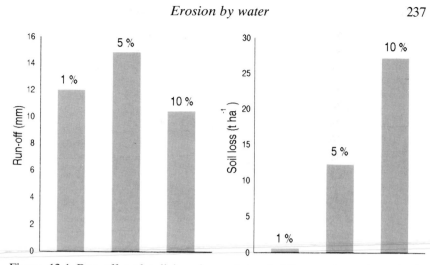

Figure 12.1 Run-off and soil loss from plots of different slopes at a site in Nigeria from 64 mm rainfall. (From Lal, R. 1975, IITA Technical Bulletin No. 1, Ibadan, Nigeria.)

12.4 Erosion by water

Water causes soil erosion mainly by (i) the impact of raindrops on the soil surface and (ii) its flow between rills and in channels downslope. It also causes landslides on steep slopes when soil overlying material of low permeability becomes saturated with water from rain or overland flow. A landslide can cause devastation when the whole soil, together with trees and buildings, slides downhill.

The erosive effect of raindrops will be understood if we consider the energy of a rainstorm. The usual formula for kinetic energy ($0.5mv^2$) can be used, where m is mass and v is velocity. The kinetic energy will be seen to vary with the square of the velocity, and measurements have shown that the bigger the raindrop the greater is its impact velocity on the ground. The range of impact velocities is from less than 1 m s^{-1} for small drops to about 9 m s^{-1} for big drops. A storm with big raindrops has therefore much greater kinetic energy than one with small drops, as shown in Figure 12.2.

Measurements are not often made of the range of drop size during a rainstorm. Empirical relationships have therefore been derived between kinetic energy and rainfall intensity, I, for example:

$$KE = 11.87 + 8.73 \log_{10}I, \qquad (12.1)$$

where I is rainfall in mm h^{-1}.

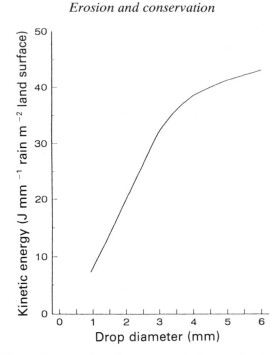

Figure 12.2 Dependence of kinetic energy of rain on the size of drops (D. Payne, unpublished).

The estimated kinetic energy of a rainstorm is often used as a measure of its erosivity. However, this over-emphasizes the effect of light rain, which has kinetic energy but might not be erosive. Erosivity of rainfall is therefore usually calculated in one of two ways to take account of the greater effectiveness of intensive rain: (i) using the kinetic energy of only that part of the rainstorm that falls above a threshold intensity, for example 25 mm per hour; and (ii) using the product of the kinetic energy of the whole storm and of the greatest 30-minute intensity; this is known as the Wischmeier index of erosivity.

The kinetic energy is dissipated when the raindrops strike the soil. One effect is to detach sand and silt grains from aggregates which then block the soil pores. Infiltration becomes less, more water runs over the surface into rills and flows downhill as run-off. Another effect of raindrops is to splash soil particles into the air. An illustration of the effect using simulated rainfall in the laboratory is given in Figure 12.3. As will be seen, on a flat surface the splash was about the same in all directions, whereas on a sloping surface the splash was greater downslope than upslope. Up to about 30° for the soil used in this experiment, the

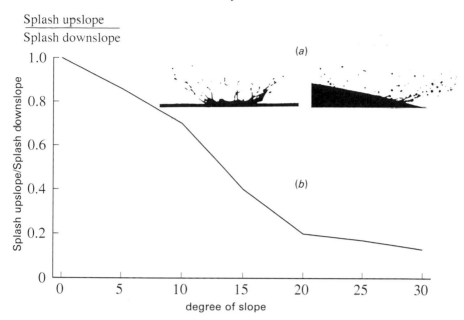

Figure 12.3 The effect of slope on splash by raindrop impact; (*a*) splash on the flat and on a 15° slope shown by high speed photography 9 milliseconds after impact; (*b*) ratio of mass of soil particles splashed upslope to mass splashed downslope in a laboratory experiment in relation to slope. (From Ghadiri, H., unpublished.)

difference became greater the steeper the slope. Splash is one of the dominant causes of erosion, yet it can pass unobserved until it is realized that the soil has become shallow towards the top of a slope.

Effects of run-off

The detachment of sand and silt grains and soil splash lower the rate of infiltration of water. Water runs off the surface if the rate of infiltration falls below the rate of rainfall (Section 6.4) and tends to collect in channels. Initially these may be natural depressions, tracks made by man or animals, or wheel marks or furrows made by farm machinery. The deeper the water in these channels the faster the flow, turbulence increases and more soil particles are detached by the flowing water and carried down slope. When the channels are sufficiently shallow to be covered over by cultivation they are called *rills*. They are called *gullies* when they are too deep to cover by normal cultivations.

Gullies can also be started by the sub-surface flow of water creating tunnels or pipes. This can occur on slopes where the subsoil has low permeability, one cause being a B horizon with a high clay content: the collapse of the overlying soil creates a gully. Once a gully is formed it can quickly become deeper and wider because of the scouring action of water in turbulent flow. As the gully deepens, the upslope head becomes unstable and collapses; the retreat of the headwall lengthens the gully. In several parts of the world gullies several metres deep have formed where there has been a combination of intense storms and erodible soil overlying unconsolidated or weathered rock.

Loss of nutrients

Soil which has been eroded becomes shallower; several experiments have shown that this results in lower crop yields. The example in Figure 12.4 is of maize yields in Georgia, USA. One cause of the decrease of crop yield is loss of nutrients in topsoil and run-off.

In one experiment in West Africa, 55 t ha^{-1} of soil was eroded over a period of 5 years. The soil that was lost contained 1200 kg organic matter, 75 kg nitrogen, 16 kg phosphorus and 34 kg exchangeable cations. Loss of topsoil, as in this example, removes that part of the soil containing the highest concentration of nutrients. From erosion plots in northern Nigeria the run-off contained annually, according to the cultivations used, between 8 and 21 kg of nitrogen per hectare; greater

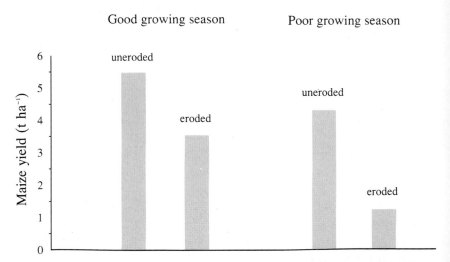

Figure 12.4 Maize yields in Georgia, USA, were decreased by erosion. (From Langdale, G.W. *et al.*, 1979. *Journal of Soil and Water Conservation*, **34**, 226.)

losses have occurred when a rainstorm followed shortly after application of fertilizer.

Nutrients in eroded soil and run-off make a significant contribution to the total nutrient input to rivers.

Effect of soil and land characteristics

The amount of erosion at any site depends on the erosivity of the rainfall, the erodibility of the soil, the characteristics of the land, and its use and management. These factors were combined by workers in the United States in an empirical equation known as the Universal Soil Loss Equation (USLE):

$$A = RKLSCP, \tag{12.2}$$

where

A = annual soil loss;
R = erosivity of rain;
K = soil erodibility factor;
L and S are length and angle of slope factors, respectively;
C = crop management factor;
P = soil conservation factor.

The reader is referred to the bibliography at the end of the book for calculations using this equation and for its limitations.

Of the six factors in the USLE, erosivity of rain has been described above. The main points about the other factors are as follows.

1.   The erodibility of soil, K, is high if it contains particles which are easily detached by raindrop impact and are transported by surface run-off. The strength and size of soil aggregates are therefore important (Section 2.2). These properties can be measured in the laboratory, for example, by shaking the soil on sieves of different mesh size under water. For use in the USLE the value of K is obtained from the loss of soil per unit of rainfall erosivity, R, using a standard field plot 22.1 m long on a 9% slope which is kept bare of vegetation and ploughed up- and downslope.

2. The angle (or percent) of slope affects both raindrop splash and the velocity of water moving over the surface, the effect of both becoming greater as the slope increases. The volume of run-off increases with the length of slope and the soil loss becomes greater.

3. Crops and natural vegetation provide a cover, which protects the

soil from the direct impact of raindrops. Permanent grass and trees provide cover throughout the year, but for annual crops the cover is present only during the growing period and when the stubble and other residues remain.

4. There are various conservation measures to reduce erosion. They include terraces, construction of protective banks along the contour, cropping in strips along the contour, or maintaining grass strips at intervals downslope. They are discussed in Section 12.8.

The acceptable limit of soil loss

Equation 12.2 can be used to predict the amount of soil that will be lost from a field when values of R, K, L, S, C and P are known. Its main purpose, however, is to provide a means of selecting appropriate conservation practices that keep the loss of soil within acceptable limits. The question then arises of the amount of soil loss that is acceptable.

As erosion is a natural process it cannot be entirely prevented. An acceptable rate, also known as the soil loss tolerance value (T), might be the rate of soil formation, so that the soil profile does not become shallower with time. The rate of soil formation can be estimated from soil depth and the period of formation, for example, the period that started with the retreat of the glaciers at the end of the last Ice Age. The estimated rates vary because of the effects of temperature, moisture and nature of the rock on the process of rock weathering. For temperate climates a formation of about 1 t ha^{-1} a^{-1} is thought to be a useful average when the subsoil is forming from an unconsolidated rock. For consolidated materials such as hard rock the rate is much less.

A much higher value of T of 11 t ha^{-1} a^{-1} has been proposed for soils deeper than 2 m which can form topsoil from medium-textured, permeable material under good management. The aim is to maintain soil fertility for 25 years and probably longer. Whatever value of T is assumed, and there is no agreement as to what the value should be, erosion can be considered to be under control if crop yields are maintained without the need for higher rates of application of fertilizer or irrigation water.

12.5 Erosion by wind

Wind causes erosion of dry, bare soil. The two factors that determine its extent are wind speed and the size of the soil particles and aggregates.

Experiments in wind tunnels have shown that the carrying capacity of wind for soil particles increases greatly with wind speed and decreases with particle and aggregate size. From these experiments:

$$c \propto v^3 \cdot d^{1/2}$$

where c is the quantity of soil removed, v is the wind speed and d is the particle diameter. Under field conditions the wind speed is commonly measured 2 m above ground level. Near the soil surface the speed is less because of frictional resistance; also, the flow of air is broken up into eddies by stones and aggregates.

The eddies create complex physical conditions at the soil surface, which are difficult to analyse. It has, however, been observed that soil particles are transported in three different ways.

1. Coarse particles (diameter above about 0.5 mm) are rolled by the wind over the soil surface. The upper limit of particle size moved in this way is about 2 mm. This kind of transport is called *soil creep*.

2. Particles of about 0.05–0.5 mm diameter are moved by a process known as *saltation*. Eddies of wind at the soil surface lift the particles into the air. The particles are too heavy to remain in suspension and after being carried for a few centimetres they fall back to the soil surface. They may bounce back into the air, roll coarse particles over the surface (soil creep) or fling into the air smaller particles which remain in suspension. Saltation causes the transport of particles of a wide range of size and is considered to be the primary process in wind erosion.

3. Particles smaller than about 0.05 mm diameter can be carried by air over large distances. It is these small particles that form dust storms which occur during droughts in semi-arid regions and are a regular feature in countries close to deserts, for example, the haboob in West Africa.

The removal of fine fractions of soil leaves the surface dominated by coarse sand, gravel or stones. The fine fractions are deposited elsewhere and become parent material for new soil. These new soils, being derived from wind-blown material, are particularly susceptible to wind erosion.

The amount of wind erosion to be expected from an area of land can be calculated from an empirical equation analogous to Equation 12.2:

$$A = f(I, K, C, L, V) \tag{12.3}$$

where

A = soil loss;
I = index of soil erodibility based on wind-tunnel tests or soil texture;
K = soil surface roughness, including the height and spacing of ridges;
C = climatic factor, including mean wind speed, rainfall and evapotranspiration;
L = length of unsheltered land downwind;
V = index of vegetative cover.

The equation can be used to determine the required conservation measures. Aside from its practical value, the equation is a useful summary of the factors that determine the amount of wind erosion.

12.6 Physical principles

Equations 12.2 and 12.3 were developed from the results of field experiments and provide a practical means of selecting appropriate conservation measures. Being empirical, they are strictly applicable only to the conditions that existed in the experiments.

Attempts have been made to base equations on physical principles. The processes in water erosion are (i) detachment of soil particles by the impact of raindrops and by run-off, (ii) transport by raindrop impact (splash) and (iii) transport by water flowing over the surface between rills and down rills. The processes are inter-related: if detachment exceeds the carrying capacity of the water, soil particles are deposited, and if it is less more particles will be detached. Equations describing these processes in physical terms are being developed which provide reasonably accurate predictions of annual soil loss.

12.7 Causes of accelerated erosion

Man's activities have increased erosion because increasing populations have required more wood for fuel and as a building material, more land to produce food crops, and have kept higher densities of cattle, sheep and goats.

The main causes of accelerated erosion are listed in Table 12.2. Whether or not there is erosion depends on the various factors given in Equations 12.2 and 12.3. Destruction of woodlands, for example, does not cause erosion on gentle slopes where rainfall is not erosive, but it

Table 12.2. *Main causes of soil erosion induced by man*

1. Destruction of woodlands: removal of the vegetative cover exposes the soil to the erosive effect of rainfall and may cause water erosion especially on steep slopes.
2. Cultivation of grasslands: water erosion occurs, as in (1) above, and in a climate which is seasonally arid, or if there is drought, the soil may blow especially if it is itself of wind-blown origin.
3. Cultivations: they leave the soil bare of a vegetative cover before there is a crop canopy and after harvest; when the soil is bare it may be subject to water or wind erosion.
4. Cultivation of steep slopes: exacerbated by cultivating up and down the slope which creates channels for run-off and may lead to the formation of gullies.
5. Intensive grazing by domestic animals: may destroy the vegetative cover and lead to wind or water erosion; goats on hill sides and herds of cattle near water-holes are particularly destructive.
6. Paths and animal tracks: collect run-off because of reduced infiltration and may lead to rill and gully formation.
7. New roads: increased run-off from roads and embankments may cause rills and gullies to form.
8. Disturbance by mining and other activities which leave the soil unprotected.

can be devastating on steep slopes with erosive rain. Similarly, cultivation of steep slopes can be made safe by the construction of bench terraces on the contour, as was done in Indonesia, Peru, the Philippines and other countries several centuries ago. Intensive grazing is safe if the vegetation is not destroyed, but if coupled with drought even non-intensive grazing by domestic animals can lead to erosion.

12.8 Soil conservation

The safe way to protect soil is not to expose it directly to wind or rain, but if arable crops are to be grown this is not practicable. Several methods have therefore been devised to protect soil against erosion (Table 12.3). The appropriate methods depend on local circumstances, including the factors given in Equations 12.2 and 12.3 and various economic, social and political conditions such as the availability of new land, land tenure, government subsidies, organization of local communities and availability of labour and farm implements. General principles will be illustrated with three examples.

1. In many countries in the humid tropics the traditional method has been for the farmer to clear a small patch of land, retaining large trees, and cropping it for two or three years before allowing it to revert to forest for ten years or longer. The soil is protected from erosion by the

Table 12.3. *Methods of conserving soil against erosion*

Biological: various ways of maintaining a cover of vegetation during the periods of high erosion risk

– good crop management
– use of rotations
– cover crops to stabilize slopes
– strip planting
– mulching with stubble and weeds
– correct stocking rate on pastures
– use of trees and hedges as windbreaks

Cultivations: use of ordinary farm implements to prepare land for an arable crop
– contour ploughing
– use of graded furrows
– minimum tillage

Mechanical protection: various forms of terrace, which are semi-permanent
– graded channel
– absorption terrace
– bench terrace
– irrigation terrace

use of small plots, the retention of large trees and the stabilization of soil structure by the organic matter which accumulates during the forest fallow. The demand for more land to grow food for an increasing population has led to shorter fallows and in some areas to continuous cultivation of large plots from which all trees have been removed. The highly erosive rain of the tropics has caused serious erosion.

Experiments have shown that erosion can often be prevented by the application of a mulch of weeds or straw and zero tillage. The mulch of organic matter maintains the infiltration rate of water by protecting the soil surface from raindrop impact, stabilizing the soil aggregates and increasing the formation of pores by the soil fauna. With zero tillage seed is drilled directly into the stubble of the previous crop, which minimizes the exposure of bare soil to rainfall.

2. The grasslands of Africa, Australia and North America are subject to erosion by wind or rain, or both, when they are cultivated. Where the land slopes gently, ploughing on the contour may be a sufficient means to control water erosion. If there is likelihood of run-off, low banks (bunds) are aligned just off the contour. Run-off from the cultivated land accumulates in grassed furrows behind the banks and runs slowly to the side of the field where it spreads over a grassed waterway before running into a stream bed. Using a ridge-and-furrow system of cultiva-

tion the same objective is achieved if the ridges run just off the contour. Where there is need to conserve water, the bunds or furrows are aligned on the contour; this stops downhill flow and allows the water to infiltrate the soil. The area between two bunds is known as an absorption terrace. Alternating strips of arable crops with a ground cover crop such as grass also protect the soil against run-off.

Where soils are subject to wind erosion, the stubble from a cereal crop, and often the straw, is left on the soil until the next crop is sown, a technique known as stubble mulching. Another method is for crops to be grown in strips at right angles to the prevailing wind between strips of tall grass, or a similar crop, which act as windbreaks. Field hedges and strips of trees also act as windbreaks. The area protected by a windbreak is fairly small; depending on wind velocity and erodibility of the soil it may extend upwind about 5 times the height of the hedge and downwind about 20 times the height.

3. Erosion is always a serious hazard on steeply sloping land in a humid climate, and conservation measures may not be practicable. Bench terracing requires stones for the walls and the availability of sufficient labour, conditions which make it expensive. If the hill sides have a cover of trees, ideally they should be left. The demand for wood as fuel has, however, resulted in the cover of trees being removed in some areas, as in parts of Nepal, and serious erosion has occurred. It seems that where there is community care of the trees, replanting is effective.

Social and economic factors

The demand for wood as fuel illustrates the need to take a broad view of conservation measures. The wood may cost only the labour of collection, and even if it has to be bought, it is usually much cheaper than the equivalent of electricity, bottled gas or kerosene (paraffin). Where the erosion hazard is particularly severe a subsidy on another fuel might therefore be the best and cheapest way to conserve the soil.

As has been mentioned earlier, the main cause of accelerated erosion has been the demand by an increasing population for more food and fuel. The easiest and cheapest solution, in the short term, is to bring more land into cultivation and to take wood from existing forests. Often this is done without any prior planning or consideration being given to the need for conservation measures. Increasingly, new land is surveyed before it is developed, and this should include assessing the risk of erosion. If the risk is unacceptably high and conservation measures are too expensive or impractical, the land should be left undisturbed. To provide the required food and wood under these circumstances, more intensive use of existing lands has to be considered using systems to maintain their fertility (Chapter 13).

Conservation measures are in the interests of both the individual farmer and of communities. Run-off from one field or plot may cause a gully to form lower down the hill on another farmer's field. Increased run-off and suspended particles can affect a whole river catchment. The support and cooperation of local communities are therefore needed, and planning of conservation measures should be based on river catchments (watersheds).

This chapter has described the effects of soil erosion on land used for agriculture. It should also be mentioned that intensive tourism in uplands and national parks involves similar processes and is an area of public concern.

12.9 Summary

Erosion of soil is a natural process which is accelerated by the removal of the natural cover of vegetation, cultivations, and overgrazing. Because it is irreversible it is generally regarded as the most serious problem of soil degradation. The effects are not limited to lower crop yields on eroded soils; there are also such harmful effects as silting up of dams and increased flooding in lower parts of the catchment.

The various factors that contribute to water and wind erosion of soils have been combined in empirical equations that are used to design conservation measures. Wind erosion is most severe in dry regions, especially those close to deserts. Water erosion depends to a large extent on the intensity of rainstorms and for this reason is severe in the humid and semi-arid tropics.

Control measures are possible. Their implementation requires the cooperation of people who work the land; financial costs have to be met, and expertise is needed in planning the required means of conservation.

13

Soils in the environment: problems and solutions

13.1 Introduction

Soils are needed to grow crops for food, animal fodder and fibre, trees for fuel and timber, and to support natural woodlands and grasslands. As the human population has increased new land has been brought into cultivation. This trend has ended in many parts of the world where all that remains is land that is either too difficult or uneconomic to cultivate. Conservation of natural ecosystems, the importance of which is being increasingly accepted, further restricts the expansion of agriculture.

Prediction of the growth of the human population is that between 1991 and 2001 the population will have risen from 5.4 billion to 6.4 billion and will have reached 10 billion some time in the second half of the twenty-first century. Much of this increase will be in the Third World. One requirement will be to manage soils so as to provide this increased population with food and with wood for fuel. This raises the question whether more intensive systems of management can be employed which, at the same time, are sustainable in environments where soils are fragile, the rains are unreliable or erosive, and the essential requirements for development are sparse.

There are different problems in countries elsewhere, especially in Europe and North America. Overproduction of food provides some justification for less intensive farming systems, with land being taken out of cultivation, for example by the set-aside scheme in the UK and the soil bank in the USA. Overproduction has led to criticism of the use of fertilizers and pesticides, and the advocacy of less intensive systems such as organic farming. There is also public concern about the quality of food, its methods of production and the quality of drinking water. One question that should concern us is how serious the hazards are from the use of fertilizers and pesticides compared with the benefits they provide.

In addition to providing the means whereby crops can be grown, soil is an important component of the terrestrial biosphere, which is that part of the land where life exists. It buffers the flow of water and solutes between the atmosphere and rivers; it also provides terrestrial vegetation with physical support, water and nutrients, and soil and vegetation together provide a mechanism for buffering the changes of carbon dioxide and other gases in the atmosphere (Table 1.1). Only recently have we begun fully to understand the importance of soils in the biosphere.

These three issues can be re-stated as referring to the appropriate use of soils and may be summarized as follows.

1. It is important to sustain the fertility of soils when their use becomes more intensive through the requirement for higher agricultural production.
2. The quality of food and water supplies should not be adversely affected by intensive methods of production.
3. Natural ecosystems should, as far as possible, be preserved.

It follows that as there is only a limited area of land on the Earth, the soil resource needs to be carefully husbanded if these requirements are to be met. It is therefore necessary to ensure that the land is used appropriately, whether for various forms of production, for example, for crops, pastures or trees, or whether it should be left undisturbed. Roads, factories, houses, airports and recreational amenities also require land. Some of the implications are considered in this chapter, but a general point will be made first.

A wide range of soils, climates, and social, economic and political conditions need to be taken into account when considering a change of land use. Unless there is good local experience, or experimentation where agricultural development is intended, caution is necessary. It should always be remembered that science is transferable, but technology is not necessarily so.

This chapter has three related themes. The first three sections deal with problems that are commonly associated with the use of fertilizers and pesticides in intensive agriculture. These are followed by discussion of the effects of intensification on soil, and of drought as a naturally occurring problem. In Sections 13.7–13.10 there is consideration of the problem of managing soils to meet the requirements of an increasing world population and, at the same time, maintaining soil fertility. The final section summarizes the three issues referred to above.

We start by discussing nitrate, its sources and as a health risk, and then as a factor in the eutrophication of water.

13.2 Nitrate, its sources and as a health hazard

Nitrate is present in most soils, the three sources being: (i) the microbial breakdown of soil organic matter, organic manures and plant residues; (ii) fertilizers which add nitrate, and that formed by the microbial oxidation of NH_4^+ from ammonium fertilizers or urea; and (iii) additions from the atmosphere (Figure 13.1).

The NO_3^- ion, being negatively charged, is not adsorbed by most soils. It remains in the soil solution until it is either taken up by plant roots,

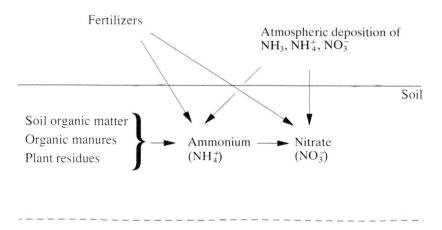

Figure 13.1 The sources of soil nitrate.

leached out of the soil in drainage water, or denitrified. It is leached out in drainage water whenever there is a sufficient excess of rainfall, as may occur during the growing season in the wet tropics and in the winter in temperate regions unless the soil is frozen. In the United Kingdom through-drainage is greatest in the wetter parts of the north and west, but occurs in all parts in most winters.

The concentration of nitrate in drainage water depends on the volume of through-drainage and the amount of nitrate that is available to be leached. High concentrations occur in eastern England where there is little through-drainage. Loss of nitrate is undesirable because (i) it represents an economic loss to the farmer, (ii) in drinking water it is considered to be a health hazard, and (iii) it may cause eutrophication (see Section 13.3).

The health risk

A hazard to health from nitrate was first recognized in Iowa in 1945. The illness of two infants was attributed to nitrate concentrations of 90–140 mg NO_3-N l^{-1} in the water used to make their feed. The water came from shallow wells and was contaminated with coliform bacteria, indicators of faecal pollution. The illness is known as methaemoglobinaemia or blue baby syndrome. It is not caused directly by nitrate but by nitrite, which can be produced in the gastro-intestinal tract by nitrate-reducing organisms. The reaction of nitrite with haemoglobin prevents the blood from transporting oxygen; this gives the skin a blue colour. In the United Kingdom the last confirmed case of methaemoglobinaemia was in 1972 and the last reported death was in 1950.

It has also been suggested that the risk of gastric cancer is increased by a high intake of nitrate in drinking water. There is evidence that after nitrate is reduced to nitrite in the gastro-intestinal tract, nitrosamines can form by the reaction of nitrosyl, a reduction product from nitrite with amines from cooked foods containing protein. Some nitrosamines have been shown to form gastric tumours in experimental animals. It has, however, been shown that the incidence of gastric cancer is less in East Anglia, where the nitrate concentration in water is higher than elsewhere in the United Kingdom. Other studies have found no evidence that high nitrate intake leads to gastric cancer, at least for populations on an adequate diet.

Nitrate analyses of drinking water are reported as concentrations of nitrate (NO_3) in milligrams per litre, or as the amount of nitrogen present as nitrate (NO_3-N) in milligrams per litre, and various standards have been introduced. The most recent from the European Commission set the 'maximum admissible level' at 50 mg $NO_3 \, l^{-1}$ (11.3 mg $NO_3-N \, l^{-1}$), and a 'guide level' at 25 mg $NO_3 \, l^{-1}$ (5.7 mg $NO_3-N \, l-1$). The health risk from nitrate in drinking water free of faecal contamination is very small, although young infants are most at risk and need protection.

The hazard to the health of humans from nitrate in drinking water is an emotive issue, with strongly expressed views influencing public opinion. The scientific and medical evidence is that the problem has occurred mainly from shallow wells contaminated with microorganisms of faecal origin.

Nitrate and nitrite are also present in the human diet and in animal feeds. High nitrate concentrations are found in leafy vegetables especially if excessive amounts of nitrogen have been added in fertilizers or manures. During transport and storage part of the nitrate is reduced to nitrite. Another source of nitrite is certain meats and cheese to which it has been added as a curing agent. About half of the total intake of nitrate and most of the nitrite are ingested from food.

Nitrate is a particular hazard to cattle and sheep because the rumen of these animals provides an environment conducive to its reduction to nitrite. Intake of nitrate from water and feeds has caused nitrite poisoning of livestock in the USA for many years.

Agricultural sources of nitrate

It has been shown that cereal crops leave very little nitrate in the soil by the time of harvest unless the crop has failed or fertilizer nitrogen has been grossly overused; grass crops that are cut and have the herbage removed also leave very little nitrate. Potatoes and other vegetable crops can leave nitrate in the soil because they tend to be heavily fertilized, are often irrigated and most are shallow-rooted; they also leave crop residues with high contents of nitrogen. Even if no nitrate is left in the soil at harvest, some is produced by mineralization and nitrification of soil organic matter, crop residues, and organic manures if these are used. The production of nitrate continues during late summer and, depending on the temperature, may continue over the winter. In these periods there is little uptake of nitrate by a crop.

The problems caused by the leaching of nitrate are therefore not caused directly by nitrogen fertilizers if (i) the fertilizer is not applied in the autumn, (ii) the recommended amount is used (iii) fertilizer is

Table 13.1. *Leaching of nitrate and concentration in drainage water*

Leaching loss (kg NO$_3$-N ha^{-1})	Drainage (mm)	Concentration (mg NO$_3$-N per l)
28	250	11.2
100	250	40
28	500	5.6

Notes
1. The Table shows the dependence of the nitrate concentration in drainage water on both the amount of nitrate leached and the volume of drainage.
2. The concentrations are average values assuming that all the nitrate is present in all the drainage water.
3. Drainage of 250 mm is equal to 25×10^5 l per ha.
4. The 'maximum admissible' concentration in water for human consumption is 11.3 mg NO$_3$-N l^{-1} in European Community countries.

applied when it is needed by the crop, and (iv) there is no crop failure. Nitrogen fertilizers can, however, have an indirect effect by increasing the amount of crop residues and soil organic matter, although the main problem is mineralization of organically-held nitrogen in the soil after crop harvest. Mineralization of soil organic matter also occurs when old grassland and woodland are brought into cultivation; when the soil is cultivated in the autumn it has been shown that the soil can lose by leaching 100 kg NO$_3$-N ha^{-1}, or more, during the following winter.

As mentioned above, the concentration of nitrate in drainage water depends on the amount leached and the volume of drainage. As shown in Table 13.1, for a drainage volume of 250 mm, which is about the value in central England, the maximum admissible concentration of 11.3 mg NO$_3$-N l^{-1} is almost reached with a leaching loss of 28 kg NO$_3$-N ha^{-1}. When compared with the total nitrogen content of soil profiles of the order of 3000 to 20 000 kg ha^{-1} it will be seen that the required standard is exceeded if a very small part of this nitrogen is converted to nitrate and leached.

13.3 Eutrophication of surface waters

The enrichment of lake, river and sea waters with nutrients that increase the growth of aquatic plants is known as eutrophication. Strictly, it is a natural process, which can be accelerated by human activities causing discharges of industrial waste waters, sewage effluent, run-off and leaching from heavily fertilized or manured agricultural land. The best known feature of eutrophication is the production of algal blooms.

There has been some uncertainty as to whether algal blooms result from increased concentrations of nitrate or phosphate or from some other cause. It is now commonly accepted that algal growth in fresh waters is generally restricted by the phosphate concentration whereas in marine waters it is restricted by the nitrate concentration. In fresh waters the nitrate concentration might, however, influence the kinds of algae that grow, some of which cause taints in drinking water or are toxic to animals.

Because phosphate is adsorbed by soil (see phosphate fertilizers in Section 8.5), the concentration in drainage water from soils is usually very low, even where phosphate fertilizers have been used. Typical concentrations are about 2×10^{-6} M, that is, 0.06 mg P l^{-1} (Table 8.11). Organically held phosphorus is more soluble, and there is concern over its leaching into surface waters, giving concentrations of 1 mg P l^{-1} or more, when large amounts of cattle and pig slurries are applied to sandy soil, as in the Netherlands. Eroded soil is also a source of phosphorus and its solubility, although normally low, increases under the anaerobic conditions that can exist on river beds and lake bottoms. Phosphorus also enters fresh waters in industrial and urban waste liquids.

13.4 Pesticides

Pesticides are chemicals that are used to control the populations of harmful organisms. An early example is Bordeaux mixture, made by mixing solutions of copper sulphate and calcium hydroxide, which has long been used in Europe to protect grape vines against fungal attack and is still used to control potato blight caused by the fungus *Phytophthora infestans*.

In the 1930s the chemical known as 2,4-D (2,4-dichlorophenoxyacetic acid) was discovered to be a selective herbicide, killing broad-leaved weeds in cereal crops. At about the same time DDT (dichlorodiphenyltrichloroethane) was found to kill a wide range of insects. A large number of pesticides, nearly all of which are synthetic organic chemicals, are now used to control fungi, weeds, insects, nematodes, mites, slugs and other crop pests. Large-scale application is limited to probably less than 50 chemicals; several of those that have been most commonly used are listed in Table 13.2.

The use of pesticides is commonly criticized, but they contribute

Table 13.2. *Some commonly used pesticides*

Insecticides	Chlorinated hydrocarbons: aldrin[a], dieldrin[a], DDT[a], lindane
	Organophosphorus compounds: malathion, fenitrathion, phorate, bromophos
	Carbamate compounds: carbaryl, carbosulfan
Herbicides	Phenoxyacetic acids: 2,4-D, 2,4,5-T[a], MCPA, mecoprop
	Triazines: atrazine, simazine
	Phenylureas: fluometuron, linuron
	Bipyridyls: diquat, paraquat
	Glycines: glyphosate
	Thiolcarbamates: EPTC
Fungicides	Dithiocarbamates: mancozeb
	Imidazoles: prochloraz
	Chlorinated aromatics: quintozene
	Benzimidazoles: benomyl
	Oxathins: carboxin

[a]Not used for agricultural purposes in the UK.

substantially to the increased production of food which has been achieved in recent years. Their manufacture and use does, however, entail some hazards to the environment. Most publicized are the effects on fish and birds. The case of DDT is referred to below; most other problems have been due to discharges from industrial plant, erosion of soil carrying pesticides into surface waters, surface run-off from land, and gross overuse of pesticides. The result is concern that soils, food crops and drinking water may have concentrations of chemicals that endanger human health.

There are some reports that pesticides affect the balance between soil organisms; for example, DDT kills species of Collembola (Section 5.3), and carbaryl reduces the number of bacteria that hydrolyse cellulose. There is, however, no evidence of any effect on soil fertility when pesticides are applied at recommended rates.

Pesticides in soils

Some pesticides are applied directly to the soil. Atrazine (chloro-ethylaminoisopropylaminotriazine) is an example of a herbicide that is applied to the soil (it is also sprayed onto weeds) to kill weed seedlings before they emerge above ground. Pesticides also reach the soil in drip from plants, as seed treatments, root dips, spray which does not contact the target organism, and in the tissues of plants and insects that have been killed. Once the pesticides are present in soil the three main factors controlling their fate are adsorption by the solid phase, decomposition and volatilization. These processes, now to be discussed, and also their water solubility, determine their mobility and persistence in soil.

Adsorption

The retention of pesticides by soils, generally referred to as adsorption, decreases their concentration in solution. The characteristics of pesticides that are generally associated with greater adsorption are (i) high molecular mass, (ii) a tendency to form positively charged ions (cations), and (iii) the presence of chemical groups that increase the affinity of the molecule for the soil surfaces.

Adsorption is generally measured by allowing the soil to react with aqueous solutions of the pesticide at a range of concentrations. At equilibrium the amount adsorbed is obtained as the difference between the amount added and the amount remaining in solution. Over a limited range of concentrations there is often a straight line relationship between the amount adsorbed, x, and the equilibrium concentration, c (Figure 13.2). The slope of the line, x/c, is known as the adsorption coefficient, K, or the distribution adsorption constant, of the pesticide by the soil used in the experiment, and is a measure of the buffer capacity. The equation $x/c = K$ is a form of the Freundlich equation (Section 4.4).

Average values of K for some commonly used pesticides (Table 13.3) vary greatly between weakly adsorbed 2,4-D and strongly adsorbed DDT and paraquat (dimethylbipyridilium dichloride). The averages are for soils with a wide range of compositions.

The content of organic carbon is often the soil characteristic that correlates best with adsorption, which can be expressed as K_{oc} using the Freundlich equation, where x is milligrams of pesticide adsorbed per kilogram of soil organic carbon (oc). Clay minerals also adsorb pesticides, the order of adsorption being smectites > illite > kaolinite, which is also the order of decreasing surface area (Table 4.2).

The mechanisms for the adsorption of pesticides by soils include:

1. Bonding by ion exchange which, for two reasons, depends on the soil pH. First, the charge on soil particles varies with pH (Section 4.1) and, secondly, pH affects the charge on the molecules of several

Table 13.3. *Values of adsorption coefficient (K) of some common pesticides*

Pesticide	K	Pesticide	K
DDT	$(1-10) \times 10^4$	atrazine	26
lindane	7–50	simazine	1–7
2,4-D	2	paraquat	200–5000

Note: The value of K varies with the properties of the soil, as explained in the text, and the equilibrium solution concentration. The large values for DDT and paraquat mean that the concentrations in the soil solution are extremely small.

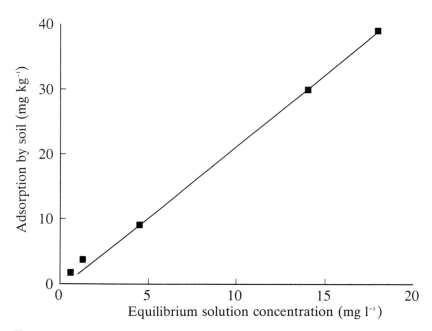

Figure 13.2 The adsorption of a pesticide, e.g. atrazine, by a soil. The slope of the line gives the adsorption coefficient, K. In this experiment K is approximately 2 mg kg^{-1}/mg l^{-1} (= 2 l kg^{-1} or 2 cm^3 g^{-1}).

pesticides. As an example, in the triazine group of herbicides, the molecule (T) becomes protonated at low pH:

$$T + H_2O \rightleftharpoons HT^+ + OH^-.$$

The cation HT^+ is adsorbed onto negatively charged surfaces of clays and organic matter.

2. Non-specific physical bonding, whereby large uncharged molecules are held by weak attractive forces.
3. Hydrogen bonding, which occurs if the pesticide contains, for example, a −NH group, the H atom acting as a bridge with an O atom on clay surfaces or an O atom in the carboxyl groups present in soil organic matter.
4. Coordination bonding, which involves a sharing of electrons between the pesticide and mineral or organic surfaces.

Weak adsorption, that is, a small value of K, indicates a high mobility of a water-soluble pesticide in soil and a large value indicates low mobility. (See Section 6.8 for the effect of adsorption on the mobility of solutes.) The value can be used to calculate the rate of transport of the pesticide to plant roots and to drainage water. Paraquat, for example, is strongly adsorbed by clays and so it is not likely to be leached out of soil (its movement over several years is very small), nor will it move to plant roots; its effect is therefore by contact with plant leaves. Water-soluble pesticides can, however, escape adsorption by being washed through cracks and pores in soil into ground water and then into aquifers; similarly, surface run-off can carry water-soluble pesticides into surface waters, e.g. rivers and lakes.

Volatilization

Pesticides have a wide range of volatility as shown by the values of vapour pressure in Table 13.4. In soil the pesticide molecules are partitioned between the vapour and liquid phases. The partition coefficient, also known as Ostwald's solubility coefficient, B, equals the concentration in liquid phase divided by concentration in the gas phase. Because diffusion is about 10^4 times as fast in gases as in water (Table 6.1), diffusion is predominantly as vapour if B is less than 10^4, and in the liquid phase if B is greater than 10^4. From the partition coefficients given in Table 13.4, it will be seen that the soil-sterilizing agent ethylene dibromide diffuses through soil as a vapour whereas simazine diffuses in the liquid phase. The principles governing diffusion of vapour are set out in Chapter 6.

Decomposition

The predominant means of decomposition are biochemical processes carried out by soil microorganisms. In addition, decomposition follows

Table 13.4. *Properties of pesticides that affect their volatility*

	Vapour pressure (mm Hg)		Solubility (μg cm^{-3})		Partition coefficient[a], solution: vapour
Ethylene dibromide	11.0	(25 °C)	4.3×10^3	(30 °C)	40
Trifluralin	1.0×10^{-4}	(25 °C)	0.58	(25 °C)	3.2×10^2
Disulfoton	1.8×10^{-4}	(20 °C)	15	(20 °C)	5.5×10^3
Dimethoate	8.5×10^{-6}	(20 °C)	3×10^4	(20 °C)	2.5×10^8
Simazine	6.1×10^{-9}	(20 °C)	5	(20 °C)	7.4×10^7

[a]The approximate value of the coefficient can be calculated as follows:
one mole of ethylene dibromide ($C_2H_2Br_2$) = 186, and in grams this occupies 22.4 l (a molar volume) at standard temperature and pressure. At a pressure of 11 mm mercury, 1 litre of air contains $\frac{11}{760} \times \frac{1}{22.4} \times 186$ g = 0.12 g of ethylene dibromide, or 120 μg cm^{-3}, giving a partition coefficient of 4300/120 = 36. In this calculation the vapour pressure was not adjusted to 0 °C, the standard temperature, and it was assumed that the vapour behaves as an ideal gas.
Source: From Graham-Bryce, I.J. 1981. In *The Chemistry of Soil Processes* (eds D.J. Greenland and M.H.B. Hayes). Wiley, Chichester; with permission.

ingestion by soil invertebrates, including earthworms, and also by plant enzymes after absorption by roots. The enzymes responsible for decomposition are similar in all organisms although their effectiveness varies greatly. The end products are carbon dioxide and water, and nitrate, sulphate or phosphate if the pesticide contains nitrogen, sulphur or phosphorus. Decomposition can also take place in sterilized soil, implying chemical degradation; hydrolysis, reduction and oxidation are the most important processes. At the soil surface most pesticides are decomposed at measurable rates by sunlight (photolysis).

Bacteria, fungi and actinomycetes are mainly responsible for the degradation of pesticides in soil, sometimes acting singly and sometimes in combination. Biodegradation of pesticides is influenced by the same processes that govern microbial activity, the organic molecules being used as an energy supply. The rate of decomposition of pesticides depends on the soil temperature and water content, their molecular structure and the effect of adsorption, which can decrease accessibility of the pesticide to microorganisms. As the structures of some pesticides are novel to microorganisms there may be a period of adaptation giving a lag phase before the more rapid phase of decomposition.

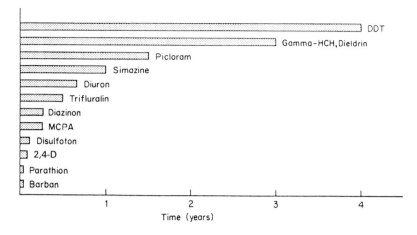

Figure 13.3 Persistence of pesticides in soils. (From Graham-Bryce, I.J., 1981. In *The Chemistry of Soil Processes* (eds. D.J. Greenland and M.H.B. Hayes). Wiley, Chichester; reprinted by permission of John Wiley & Sons Ltd., copyright 1981.)

Persistence

Pesticides are subject to adsorption, decomposition and volatilization as described above. Ideally a pesticide should remain long enough to kill the target organism and then be decomposed to harmless degradation products. As shown in Figure 13.3, 2,4-D is decomposed in soil within about one month whereas DDT can persist for several years; its principal degradation product, DDE, is also persistent. With repeated applications, concentrations of the more persistent pesticides will increase and may damage non-target organisms.

Pesticides in the environment

The three main uses of pesticides which are of benefit to man are:

1. To control weeds, fungal diseases and insect and other pests which would otherwise reduce the yield of agricultural crops.
2. To stop the spoilage of food crops during storage by controlling insect pests such as grain weevils.
3. To kill disease-carrying organisms.

Environmental hazards were first recognized with DDT, an organochlorine insecticide. It was widely used in the 1940s and 1950s to kill mosquitoes and may thereby have saved over a million lives from malaria. Three properties of DDT, however, together cause problems. First, DDT and its principal decomposition product DDE are highly persistent; secondly, it is soluble in fat and therefore accumulates in fatty and oily tissues; thirdly, it is toxic to fish and birds.

Applications of DDT which reach the soil are taken up by earthworms and insects. Concentrations increase in birds eating the earthworms and insects, and increase further in predatory birds such as the eagle. Similarly in water, DDT is taken up by plankton which are consumed by small fish, which are consumed by bigger fish and birds, the concentration of the chemical in tissues of the animals increasing with each step and causing their deaths. For this reason the use of DDT is now prohibited in many countries.

Because of the experience with DDT, thorough testing of the environmental impact of a chemical is required before it can be marketed. Based on the results of these tests the pesticides now in use are generally considered safe as long as the guidelines governing their use are followed. There is, however, public concern that they might cause cancer or act as poisons, although there is little evidence to justify these fears.

Environmental checks are made by measuring the concentrations of pesticides in water and food. In the European Community the maximum permissible concentrations in drinking water are:

1. $0.1\ \mu g\ l^{-1}$ for any individual pesticide and related products; and
2. $0.5\ \mu g\ l^{-1}$ for total pesticides and related products.

These concentrations are based on detection limits rather than on concentrations that are a risk to health. They are nevertheless mandatory.

13.5 Degradation of soil

There are several possible causes of soil degradation (Table 13.5). Some have been discussed in earlier chapters: waterlogging and compaction (Section 8.2), erosion (Chapter 12), acidification (Chapter 9), salinization and sodification (Section 8.10) and the accumulation of heavy metals and other inorganic contaminants (Chapter 10). All of these, and particularly erosion and acidification, can affect the environment in

Table 13.5. *Causes of soil degradation*

Erosion
Acidification
Salinization and sodification
Accumulation of toxic elements
Depletion of plant nutrients
Reduction of soil organic matter content
Compaction and crusting
Waterlogging, except for rice

various ways. In this section their effects on soil fertility are summarized. Also to be discussed are the effects of the depletion of plant nutrients and the importance of soil organic matter.

Accelerated soil erosion is the greatest hazard in most environments to the long-term maintenance of soil fertility. It reduces soil depth, and might remove the entire soil. Loss of topsoil means loss of the layer with most nutrients, most organic matter and the best structure for root growth. Erosion can reduce crop yields to zero, although loss of a few tonnes of soil per hectare might have no noticeable effect if the soil profile is deep. Globally, accelerated erosion is the most obvious form of soil degradation. Control measures can be expensive.

Less obvious is the effect of soil acidification. As described in Chapter 9 it is a natural process, which is accelerated by the use of fertilizers containing ammonium salts or urea, by biological nitrogen fixation, nutrient removal in crops, and deposition of acids and acid-forming substances from the atmosphere. Unless the parent material supplies basic cations at a sufficient rate, soils under natural vegetation in a humid climate become acid. The pH of a weakly buffered soil can drop from 6.0 to 4.5 in two or three years if 100 kg N ha^{-1} a^{-1} is added as ammonium sulphate, one of the most acidifying fertilizers. Soil acidity is a widespread problem which is expensive to remedy unless local sources of lime are available.

Salinity and sodicity are problems that occur particularly in semi-arid environments when crops are irrigated (Section 8.10). The problems can be prevented by using excess water and installing a drainage system. As with measures to conserve soils and rectify soil acidity described above, prevention costs money.

The contamination of soils with heavy metals is mostly from mining and manufacturing processes, the use of lead in petrol and the disposal of industrial wastes. Unlike contamination with organic substances, heavy metals remain in the soil and where contamination is severe there is no remedial treatment except to replace the contaminated soil. Severe contamination with radionuclides requires the same, extreme method of treatment. Safety checks on food and water are needed for less severe contamination.

Soil degradation can also be due to depletion of plant nutrients. Harvested crops, and animal products to a less extent, remove nutrients; for example, on average, each tonne of cereal grain removes about 20 kg N, 4 kg P, 6 kg K, 2 kg S, 1 kg Ca, 1 kg Mg and smaller amounts of the micronutrients (Table 8.1). Much larger amounts of nitrogen, sulphur, calcium and magnesium can be lost by leaching in a humid environment, and nitrogen is lost to the atmosphere by denitrification and volatilization of ammonia. Phosphate, calcium, magnesium, potassium and the micronutrients are slowly replenished if soil minerals containing these elements are present in the soil. Additions to the soil are, however, necessary if depletion of plant nutrients is to be avoided.

Organic matter is not an essential component of soils but it has important effects on soil structure and is a source of nutrients, particularly nitrogen, phosphorus and sulphur. The content of organic matter decreases during cultivation to about one half to one third of that present after a long period under grass or trees (the dynamics of soil organic matter is discussed in Section 3.4). It can be increased by application of organic manures or by putting the land down to grass for a period of years, as is done in ley farming. Many experiments have compared crop yields on plots receiving fertilizers with yields on plots with applications of organic manures or after grass has been grown. Examples of results are given in Figure 8.6 and Table 8.3. From a thorough review of experiments, the conclusion of Cooke was that economic yields may be obtained in permanent cropping systems without any special action being necessary to add organic matter by manures or growing leys, and that their use should be determined by the farming system followed, which should be profitable.

Cooke regarded his conclusion as tentative and not necessarily applicable to all soils. He stressed that organic matter might be more beneficial in harsher climates. There is further discussion of the management of soil organic matter in Section 13.8.

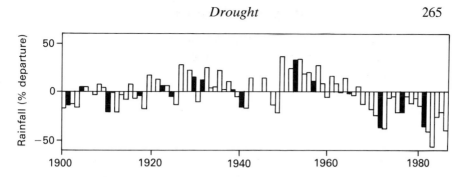

Figure 13.4 Rainfall fluctuations in the African Sahel 1901–87, expressed as per cent departure from the long-term mean. (From Nicholson, S.E., 1989. *Weather* **44,** 46; with permission of the Royal Meteorological Society.)

13.6 Drought

The requirements of plants for water are discussed in Section 7.4, where reference is made to the characteristics of rainfall which affect the supply of water to plants and to their strategies for surviving periods of water shortage.

Drought is a major problem faced by communities directly dependent on rainfall for drinking water, the growth of crops and the rearing of animals. Drought may be defined arbitrarily as a certain number of days without rain, or as a period when the rainfall is substantially below normal. For agricultural crops it is better defined as a period in which lack of water substantially reduces the growth and final yield of the crop. In a region of marginally sufficient rainfall, the effect of rainfall which is substantially below normal can be catastrophic. The problem can occur anywhere but in recent years has been particularly severe in the African Sahel.

The Sahel stretches across the southern fringe of the Sahara Desert from West Africa to Sudan. It has summer rainfall with a long-term average, according to the region, of up to about 500 mm which falls in a short summer season. The rainfall began to decrease in the 1950s and for several years since the late 1960s remained below the long-term average (Figure 13.4). The cause of the reduction of rainfall is not yet established, but seems to be linked with a change in temperature of the sea surface which itself is not yet understood. It is known that long droughts followed by periods of above-average rainfall have occurred in the past in the Sahel, suggesting fairly frequent cyclic fluctuations of rainfall.

Before the onset of the recent drought, a succession of years with good rains encouraged farmers to grow crops requiring a longer growing season, such as sorghum instead of millet, and to cultivate land where rainfall had previously been too low. The higher rainfall also improved the growth of pastures, which could then sustain a greater number of cattle. The return of drought caused crop failure and starvation of animals, the effect being exacerbated by the presence of more people and more animals.

The return of drought also raised fears that desert-like conditions would spread (desertification, also known as desertization), possibly intensified by the effects of cultivations. Wind erosion has occurred and in some regions has been severe, as might be expected where the soils have formed from wind-blown material.

Ideally, land use in the Sahel should be based on the expectation that droughts will recur, the corollary being that more intensive land use will be required in areas with a more dependable supply of water as rainfall, or by irrigation in valley bottoms.

13.7 Attaining higher crop yields

Scientific and technological developments during the past few decades have led to sufficient production, or even overproduction, of food in Europe, North America, and in a few other countries. But in much of the world higher production is needed, and will continue to be needed in the foreseeable future. In this section we refer to the possibilities, and to some of the problems that have to be overcome.

Most arable crops (wetland rice is an important exception) require similar soil conditions if they are to grow well (Table 13.6). 'Soil fertility' is the term used for these conditions and can be defined as the capacity of the soil to support the crop being grown. The individual conditions have been discussed in previous chapters. Good soil management entails providing these conditions, but there are more requirements (Table 8.5) if high crop yields are to be achieved. Some of these requirements are illustrated by the history of grain yields in England.

Until the start of the sixteenth century, the yield of wheat in England was less than 1 t ha^{-1}. Enclosure of the land and the use of fallows raised the yield to about 1.5 t ha^{-1} by the middle of the eighteenth century. By 1900 the yield had increased to over 2 t ha^{-1} owing to the use of rotations, better methods of cultivation and sowing, and the introduction of fertilizers. The yield has since increased more rapidly because of

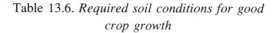

Table 13.6. *Required soil conditions for good crop growth*

Adequate soil depth
Suitable mechanical properties
High available water capacity
Aeration of the root zone
Suitable temperature
Non-limiting supply of nutrients
pH about 6
No toxic concentrations of metals, salts or herbicides

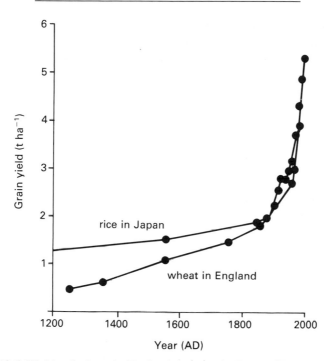

Figure 13.5 Yields of wheat in England and rice in Japan. (From Evans, L.T., 1980. *American Scientist* **68**, 388; reprinted with permission of *American Scientist*, Journal of Sigma Xi, The Scientific Society.)

better crop varieties and the use of more fertilizer and pesticides, and now exceeds 5 t ha^{-1} (Figure 13.5). Yields of other arable and grass crops have also increased substantially. In Japan the yield of rice has increased similarly to that of wheat in England.

In the Third World yields of arable crops can often also be increased substantially by appropriate cultivations, the growing of high-yielding varieties and the use of fertilizers, pesticides and irrigation. Over the past 35 years, for example, the average yield of wheat in India has increased from 0.74 to 3.5 t ha^{-1}, and in China from 0.74 to 3.04 t ha^{-1}.

The success of the so-called 'Green Revolution' depends on a package of inputs, such as were gradually introduced in the past, illustrated above for wheat in England. They include new, short-stemmed crop varieties with a high harvest index, the use of fertilizers and pesticides, and often of irrigation. Where the change of production has been successfully introduced the farmer has been able to purchase the required seed and chemicals, often using government subsidies. In many countries, however, the potential yield increases are not being attained because the required inputs either cannot be paid for or are not available, and storage and marketing of the produce are inadequate.

There can also be difficulty with the introduction of irrigation, which can be expensive and is not always successful. The package of inputs referred to above is necessary because high crop yields are needed to justify the cost of an irrigation scheme. Further, farmers have to adapt to managing irrigation, which takes time if it is to be done well. There can also be health problems due to the spread of malaria and bilharzia.

The general consensus is that the most intractable problems limiting development of agricultural productivity in the Third World are economic and political. Technical problems of soil management are being solved, but there is need for continued investigation of farming systems that are acceptable to farmers and give sustainable yields in the long term. Sustainability is discussed in Section 13.9. Next to be considered is the place of organic farming.

13.8 Organic farming

The general principle of organic farming is to create a system which relies on biological processes for the production of crops and livestock and their protection from pests and diseases. More specifically, it avoids the use of pesticides and of water-soluble fertilizers, and aims to produce food which is regarded by the consumer as safer or more healthy than that produced by conventional farming.

To meet the requirements for nitrogen organic farmers use leguminous crops, the nitrogen left in the soil supplying part of the requirements of the succeeding non-leguminous crop. Ground rock phosphate and basic slag are used instead of water-soluble phosphate fertilizers, and crushed rock (some shales are suitable) is substituted for conven-

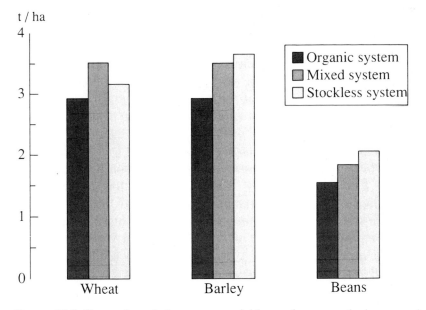

t / ha

Figure 13.6 Comparison between crop yields on farms employing organic, mixed or stockless systems (the Haughley experiment). (From Stanhill, G., 1990. *Agriculture, Ecosystems and Environment* **30,** 1.)

tional potassium fertilizers. The control of weeds, disease and insect pests is by means of agronomic practices that include the use of rotations, appropriate cultivations and the selection of crop varieties resistant to disease and insect pests. Animal manures, composted plant residues and green manures are used to maintain the level of soil organic matter.

There has been a variety of practices in organic farming which are now becoming standardized. The basic definition is, however, that it is farming without the use of soluble fertilizers or synthetic pesticides. Its economic viability depends on the higher price that customers will pay for organically grown food because yields are generally lower than from conventional farming. Two examples will be given.

The first example is of crop yields (Figure 13.6) obtained in eastern England between 1952 and 1965 on a 91 ha farm which had been split in 1941 into three separate units: organic, mixed and stockless. Dairy cattle, pigs, sheep and poultry were kept on the organic and mixed units; crop residues and animal manures produced within each unit were used on the arable crops. Fertilizers were used on the stockless and mixed units. The average yields over 14 years on the organic unit as percentage of those on the stockless unit were 93, 80 and 75 for wheat, barley and beans respectively.

A comparison between organic and conventional systems in the north-east of the USA has been reported by CAST (1980). The yields on organic farms expressed as percentages of those on conventional farms ranged between 56 and 107 with an overall average of 83. Similar comparative yields have been reported from other countries. Yields depend, however, on the level of inputs in the two systems.

The proponents of organic farming usually see it as a self-contained system which requires low external inputs to maintain soil fertility and produces food of high quality. These objectives, together with high yields, which are usually essential for profitability, are those of any good farmer. Where customers are prepared to meet the higher cost, organically grown food finds a ready market. In the Third World, low inputs are often necessary, as is the conservation of soil fertility. Yields of crops and livestock have, however, to be raised and strict adherence to the principles of organic farming, which exclude the use of soluble fertilizers and pesticides, would generally not meet this requirement.

13.9 Sustaining soil fertility

As defined earlier, soil fertility is the capacity of the soil to support the crop being grown. If any form of soil degradation occurs (see Section 13.5) the fertility will become less and might be lost altogether. In addition to the causes listed, increase of weeds, disease organisms, insect and other pests will harm the crop and should therefore be avoided.

The practices adopted by the farmer to sustain the fertility of his soil depend on his physical environment and economic circumstances. Where erosion is likely to occur, conservation measures are a priority, and where fertilizers are expensive or not available it is important to recycle nutrients and use some system of biological nitrogen fixation. Elsewhere it might be necessary to prevent the soil from becoming too acidic. In other words, there is no one method of sustaining soil fertility that is universally applicable.

Traditional methods in the UK included crop rotations, the use of farmyard manure, the growth of legumes and bare fallows. The rotation of crops controlled the populations of weeds, disease organisms and harmful insects; farmyard manure provided nutrients and improved the physical condition of the soil; legumes provided nitrogen, and bare fallows gave additional control of weeds and allowed mineral nitrogen to accumulate in the soil. A bush fallow, which is part of the rotation traditionally practised in much of the tropics, similarly controls pests, and adds nutrients to the soil surface in leaf litter and in ash when the understorey is burned. These traditional methods maintained crop yields, though at a low level, and can fairly be said to have sustained soil fertility.

Since the 1950s, crop and animal production has become more intensive in the industrialized countries. In the UK animal production has been largely transferred to the wetter west of the country leaving the eastern part devoted mainly to arable crops that are often grown continuously. There was concern in the late 1960s that one consequence would be a loss of soil fertility, particularly of soil structure on silty soils. As yet there is no evidence to indicate that crop yields are being reduced.

There is more risk from intensification in the tropics especially from erosion. Nevertheless, at one experimental site in northern Nigeria where good conservation measures were used, mechanical cultivation and the use of fertilizers and pesticides continued for several years to give crop yields about five times as high as those obtained by local farmers. A cheaper and less intensive system would also have been successful as long as it included soil conservation measures.

Agroforestry

A form of agroforestry has been the traditional method of maintaining soil fertility in the tropics. After small plots have been cropped for two or three years the land is allowed to revert to bush (forest) for a period of ten to twenty years. The rest period becomes shorter when the population increases; soil erosion often follows.

As the term is now used, agroforestry refers to a form of land use in temperate and tropical regions which helps to control soil erosion and maintain soil fertility by growing trees and shrubs in association with crops or pastures. They can be grown in a variety of ways, for example, in rotation with crops, as hedgerows separating strips of cropped land, or by retaining individual trees or groups of trees on land which is cropped or under pasture. Species that are planted in the tropics include *Leucaena leucocephala*, which fixes nitrogen; it can provide 100–200 kg N ha^{-1} a^{-1} in its prunings, which can be used as a mulch. Many other species are being investigated.

Table 13.7. *Beneficial and adverse effects of trees on soil fertility*

Beneficial effects
1. Litter protects the soil against erosion
2. Litter adds organic matter and nutrients to the soil surface (see Table 8.7)
3. Trees reduce the loss by leaching of plant nutrients
4. Some species fix nitrogen in symbiosis with bacteria

Adverse effects
1. Compete with arable and grass crops for water and nutrients
2. Can cause acidification
3. Nutrients are removed in tree harvest

Trees provide economic benefits by being a source of wood as building material and for fuel. They also have beneficial effects on soil fertility although they can also have adverse effects (Table 13.7). They can protect the soil against erosion because of (i) the ground cover of litter, and (ii) the presence of an understorey of vegetation. Uptake of nutrients, retention in their tissues and return to the soil surface in litter reduces loss by leaching. Another benefit is the fixation of nitrogen in symbiosis with *Rhizobium* by some species. They do, however, compete with arable and grass crops for light, water and nutrients; they can also cause soil acidification (Figure 9.3b).

For soil fertility to be sustained under more intensive use where there is little previous experience, soil surveys are needed, and these should include identification of possible causes of degradation (see next section). If there are hazards, control measures are needed which should be followed by regular monitoring of the soil properties most at risk.

13.10 Making the best use of soils

Where new land is to be developed, or land use is to be changed, it is now normal to make a survey of the resources. For agriculture these resources include characteristics of the total, relevant environmental conditions including climate, slope of the land surfaces, soils, hydrology, vegetation, and the people, their social organization and access to markets. The survey, known as a general purpose land evaluation, requires information obtained on the ground and is aided by aerial photography and remote sensing.

Whatever the purpose of the survey, experience and skill are required of the surveyor if it is to be successful. Only a few points are referred to

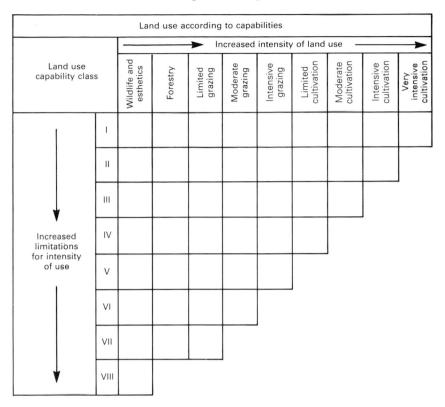

Figure 13.7 The Land Use Capability Classes as developed by the United States Soil Conservation Service.

here; the books suggested for further reading should be consulted for more thorough accounts, for example Landon (1991).

A system of land evaluation developed in the USA and widely used for agricultural development projects is based on the limitations for agricultural use. It is known as a land capability classification. The scheme shown in Figure 13.7 has eight land classes, and within each class the limitation might be erosion hazard, water excess, stoniness, salinity, shallow profile, or climatic limitations. The land capability boundaries are identified by clustering soil mapping units. An example of the successful use of land capability classes is the Canadian Land Inventory.

Land resource data can also be used to assess land suitability, a concept introduced by FAO. Whereas each capability class covers a broad range of crops at moderately high levels of management, a suit-

Table 13.8. *Some examples of successful soil surveys and land evaluation studies*

1. Agro-pastoral development in Queensland, Australia
2. Irrigated sugar cane in Swaziland
3. Village-based soya cultivation in Andhra Pradesh, India
4. Water harvesting in Israel and northern Kenya
5. National Park management, Waterton Lakes in Alberta, Canada
6. Establishing and monitoring recreational sites in Hong Kong
7. Urban subdivision design and effluent dispersal in USA
8. Establishment of a gas-pipeline network in UK

ability class relates to a particular crop, specific guidelines having been published for irrigated and rain-fed crops, and for forestry. At all scales, from an individual field, to farm, catchment and region, three levels of suitability, and non-suitable, may be established and mapped. The system is being developed further by FAO to establish agroecological zones to describe the present and potential use of the world's land resources.

Another concept is that of agrotechnology transfer, initiated in the USA and developed in the Benchmark Soils Project. The underlying assumption is that if management and crop yield are known on a soil whose properties are described, similar yields can be expected on soil with the same properties and the same management at another location, even in another country. Lower yields might indicate the need to improve management.

Before mapping new land it is first necessary to know what use is intended for the land. Only those properties are then identified and mapped that are necessary to establish the suitability of the land for its intended use. The purpose of the survey might be to use the land for some form of agriculture, or for forestry, construction of highways or buildings, recreation or nature reserves.

There have been many successful surveys, a few of which are listed in Table 13.8, to indicate the range over which they have been used. They have generally been most successful at a detailed scale and when the surveyor has had at least partial responsibility for overseeing their implementation.

13.11 Summary

The properties of soils and the processes that occur in them (Part A) influence the environment in ways that have been described in Chapters 7–12, some of which are brought together in this chapter.

Because of man's dependence on the production of crops and domestic animals it is understandable that much of our knowledge about soils has been related to agriculture. Experience and experiments have shown that some virgin soils can be cultivated safely and are productive whereas others require the addition of plant nutrients or lime; some require protection against erosion and others are best kept undisturbed. It is now realized that the soils of the Earth are a limited resource and need to be conserved. They are, however, very variable, and make generalizations about their use a trap for the unwary.

By working together scientists and farmers in some countries have shown that high yields of crops and animal products can be achieved with minimum damage to the environment. But in the past, and to the present day, man has been careless in the use of soils, especially in causing their erosion. A problem for the future is how to raise yields to sustain a larger human population without damaging the soil. There will be no one method of achieving this goal. It will be necessary to analyse the requirements region by region, taking account of the local physical, biological, social and economic circumstances. As mentioned earlier in this chapter, new land might be brought into cultivation in some parts of the world; elsewhere there is potential for higher yields through intensification. How either development should proceed has been outlined in earlier sections.

In addition to their use in agriculture, soils also support the natural ecosystems of the Earth, although these have been relatively neglected by soil scientists and deserve more intensive study. The potential for global warming due to increased concentrations of radiatively active (greenhouse) gases in the atmosphere is of current concern. Soils are both source and sink for some of these gases (Chapter 11), and they may also influence the global energy budget by producing dust in the atmosphere from wind erosion. The terrestrial environment is also affected by the deposition of acids and acid-forming substances from the atmosphere, especially as a result of the combustion of fossil fuels (Chapter 9). Soil buffers the flow of water between the atmosphere and the hydrosphere, and also reduces the effects of the acids on trees, crops and water supplies.

To summarize: there are several interactions between soils and other components of the environment. The long-term conservation of soils and maintenance of their fertility is essential for life as we know it, and will be of critical importance during the next few decades.

Suggestions for further reading

There are several useful textbooks for first year undergraduates. The following list will supplement the material in Chapters 1–6.

Brady, N.C. 1990. *The Nature and Properties of Soil*, 10th edition. Macmillan, New York.
Faniran, A. and Areola, O. 1979. *Essentials of Soil Study, with Special Reference to Tropical Areas*. Heinemann, Nairobi.
FitzPatrick, E.A. 1986. *An Introduction to Soil Science*, 2nd edition. Longman, Harlow.
McLaren, R.G. and Cameron, K.C. 1990. *Soil Science. An Introduction to the Properties and Management of New Zealand Soils*. Oxford University Press, Auckland, N.Z.
White, R.E. 1987. *Introduction to the Principles and Practice of Soil Science*, 2nd edition. Blackwell Scientific Publications, Oxford.

Three books that give a fuller treatment of parts of Chapters 7–13 are:

Harrison, R.M. (ed.) 1990. *Pollution: Causes, Effects and Control*. Royal Society of Chemistry, Cambridge, UK.
Hillel, D. 1991. *Out of the Earth*. The Free Press, New York.
Schlesinger, W.H. 1991. *Biogeochemistry. An Analysis of Global Change*. Academic Press, San Diego.

The following are useful for particular chapters.

Chapter 1

Bridges, E.M. and Davidson, D.A. (eds.) 1982. *Principles and Applications of Soil Geography*. Longman, London. See Chapter 1 by D.A. Davidson.

Chapter 2

Bohn, H.L., McNeal, B.L. and O'Connor, G.A. 1979. *Soil Chemistry*. Wiley Interscience, New York.

277

Hillel, D. 1980. *Fundamentals of Soil Physics*. Academic Press, New York.
Sposito, G. 1989. *The Chemistry of Soils*. Oxford University Press.

Chapter 3

Buol, S.W., Hole, F.D. and McCracken, R.J. 1989. *Soil Genesis and Classification*, 3rd edition. Iowa State University Press, Ames.
Ollier, C.D. 1976. *Weathering*. Longman, Harlow.

Chapter 4

Books by Bohn, McNeal and O'Connor, and by Sposito, as for Chapter 2.

Chapter 5

Alexander, M. 1977. *Introduction to Soil Microbiology*, 2nd edition. Wiley, New York.
Anderson, J.M. 1981. *Ecology for Environmental Science: Biosphere, Ecosystems and Man*. Arnold, London.
Jackson, R.M. and Raw, F. 1970. *Life in the Soil*. Arnold, London.
Paul, E.A. and Clark, F.E. 1989. *Soil Microbiology and Biochemistry*. Academic Press, San Diego.
Swift, M.H., Heal, O.W. and Anderson, J.M. 1979. *Decomposition in Terrestrial Ecosystems*. Blackwell Scientific Publications, Oxford.
Wood, M. 1989. *Soil Biology*. Blackie, Glasgow.

Chapter 6

Book by Hillel, as for Chapter 2.

Chapter 7

Harley, J.L. and Smith, S.E. 1983. *Mycorrhizal Symbiosis*. Academic Press, San Diego.
Jordan, C.F. 1985. *Nutrient Cycling in Tropical Forest Ecosystems*. Wiley, Chichester.
Likens, G.E., Bormann, F.H. *et al.*, 1977. *Biochemistry of a Forested Ecosystem*. Springer-Verlag, New York.
Marschner, H. 1986. *Mineral Nutrition of Higher Plants*. Academic Press, San Diego.
Mengel, J. and Kirkby, E.A. 1987. *Principles of Plant Nutrition*, 4th edition. International Potash Institute.

Chapter 8

Archer, J. 1985. *Crop Nutrition and Fertilizer Use*. Farming Press, Ipswich.
Briggs, D. and Courtney, F. 1985. *Agriculture and Environment. The Physical Geography of Agricultural Systems*. Longman, Harlow.

Cooke, G.W. 1982. *Fertilizing for Maximum Yield*. Granada, London.
Milthorpe, F.L. and Moorby, J. 1979. *An Introduction to Crop Physiology*, 2nd edition. Cambridge University Press.
Sanchez, P.A. 1976. *Properties and Management of Soils in the Tropics*. Wiley, New York.
Tisdale, S.L., Nelson, W.L. and Beaton, J.D. 1985. *Soil Fertility and Fertilizers*. Macmillan, New York.
Wild, A. (ed.) 1988. *Russell's Soil Conditions and Plant Growth*. Longman, Harlow.
For irrigation, see James, D.W., Hanks, R.J. and Jurinak, J.J. 1982. *Modern Irrigated Soils*. Wiley, New York.

Chapter 9

CAST, 1984. *Acid Precipitation in Relation to Agriculture, Forestry and Aquatic Biology*. Council for Agricultural Science and Technology, Ames, Iowa.
Longhurst, J.W.S. (ed.) 1989. *Acid Deposition. Sources, Effects and Controls*. British Library Technical Communications.
Mellanby, K. (ed.) 1988. *Acid Pollution, Acid Rain and the Environment*. Watt Committee Report No. 18. Elsevier, London.
Wellburn, A. 1988. *Acid Pollution and Acid Rain: The Biological Impact*. Longman, Harlow.

Chapter 10

Alloway, B.J. (ed.) 1990. *Heavy Metals in Soils*. Blackie, Glasgow.
Bar-Yosef, B., Barrow, N.J. and Goldschmidt, J. (eds.) 1989. *Inorganic Contaminants in the Vadose Zone*. Springer-Verlag, Berlin.
Davies, B.E. 1980. *Applied Soil Trace Elements*. Wiley, Chichester.
Leeper, G.W. 1978. *Managing the Heavy Metals on the Land*. Dekker, New York.
Eisenbud, M. 1987. *Environmental Radioactivity*. Academic Press, Orlando.

Chapter 11

Bolin, B., Döös, B.R. and Jäger, J. (eds.) 1986. *The Greenhouse Effect, Climate Change, and Ecosystems*. Wiley, Chichester.
Bouwman, A.F. (ed.) 1990. *Soils and the Greenhouse Effect*. Wiley, Chichester.
Houghton, J.T., Jenkins, G.J. and Ephraums, J.J. (eds.) 1990. *Intergovernmental Panel on Climate Change*. Cambridge University Press.
Warneck, P. 1988. *Chemistry of the Natural Atmosphere*. Academic Press, San Diego.
Wayne, R.P. 1991. *Chemistry of Atmospheres*, 2nd edition. Clarendon Press, Oxford.

Chapter 12

CAST, 1982. *Soil Erosion: Its Agricultural, Environmental, and Socioeconomic Implications*. Council for Agricultural Science and Technology, Ames, Iowa.

Hudson, N. 1971. *Soil Conservation*. Batsford, London.

Morgan, R.P.C. 1986. *Soil Erosion and Conservation*. Longman, Harlow.

Troeh, F.R., Hobbs, J.A. and Donahue, R.L. 1980. *Soil and Water Conservation for Productivity and Environmental Protection*. Prentice Hall, New Jersey.

Chapter 13

Attiwell, P.M. and Leeper, G.W. 1987. *Forest Soils and Nutrient Cycles*. Melbourne University Press.

CAST, 1980. *Organic and Conventional Farming Compared*. Council for Agricultural Science and Technology, Ames, Iowa.

Cooke, G.W. 1967. *The Control of Soil Fertility*. Crosby Lockwood, London.

Dent, D. and Young, A. 1981. *Soil Survey and Land Evaluation*. Allen and Unwin, London.

Lal, R. 1987. *Tropical Ecology and Physical Edaphology*. Wiley, Chichester.

Lampkin, N. 1990. *Organic Farming*. Farming Press, Ipswich.

Landon, J.R. (ed.) 1991. *Booker Tropical Soil Manual*. Longman, Harlow.

Woodwell, G.M. 1990. *The Earth in Transition. Patterns and Processes of Biotic Impoverishment*. Cambridge University Press.

Young, A. 1989. *Agroforestry for Soil Conservation*. CAB International, Wallingford, UK.

Books for soil analysis

MAFF 1986. *The Analysis of Agricultural Materials*. MAFF Reference Book RB 427. HMSO, London.

Methods of Soil Analysis Part I (ed. A.L. Page), 1982 (*Chemical and Microbiological Methods*); Part II (ed. A. Klute), 1986 (*Physical and Mineralogical Methods*). American Society of Agronomy, Madison, Wisconsin. This is the reference publication on methods.

Rowell, D.L. 1993. *Soil Science: Methods and Applications*. Longman, Harlow.

Walsh, L.M. and Beaton, J.D. 1980. *Soil Testing and Plant Analysis*. Soil Science Society of America, Madison, Wisconsin.

Index

281